The Digital Playbook

Dirk Slama
AIoT User Group Editor-in-Chief

Tanja Rückert •
Sebastian Thrun • Ulrich Homann • Heiner Lasi
Editors

The Digital Playbook

A Practitioner's Guide to Smart, Connected Products and Solutions with AIoT

Sponsored by:

Editors
Dirk Slama
Robert Bosch GmbH
Ferdinand Steinbeis Institute
Berlin, Germany

Sebastian Thrun
Udacity Inc
Mountain View, CA, USA

Heiner Lasi
Ferdinand-Steinbeis-Institut
Stuttgart, Germany

Tanja Rückert
Robert Bosch GmbH
Stuttgart, Germany

Ulrich Homann
Microsoft (United States)
Redmond, WA, USA

The open access publication of this book has been published with the support of the Ferdinand-Steinbeis-Institute.

ISBN 978-3-030-88220-4   ISBN 978-3-030-88221-1   (eBook)
https://doi.org/10.1007/978-3-030-88221-1

Preface

In today's digital world, no stone is left unturned. Even companies with traditional businesses around physical products and equipment are now at the forefront of the digital revolution. Product companies, OEMs, manufacturers, and equipment operators are continuously digitalizing to make their businesses more customer centric, automate repetitive tasks, and even create new digital businesses.

Smart connected products are the next step in the evolution of physical products, enabling new digital features and services. Smart connected solutions help equipment operators to continuously improve Overall Equipment Effectiveness (OEE).

Data, Artificial Intelligence (AI), and software are the key enablers. The Internet of Things (IoT) is providing the connectivity between physical products and equipment with the cloud. AI in combination with IoT-generated data creates new insights and predictions. It is even enabling autonomous vehicles, products, and equipment. This is a huge step forward from the basic IoT applications of the past.

AIoT – the Artificial Intelligence of Things – is the next paradigm shift. AIoT provides the foundation for smart connected product and solutions. By combining AI and IoT, AIoT has the potential to profoundly change the way in which we will do business in the future.

However, in order to get there, it is vital to ensure that the many delicate little seedlings and pilot projects that we are seeing across the industry are transformed into real businesses and solutions at scale – either generating additional revenue, or saving time and money. This will require us to rethink how we are approaching the technical innovation potential of AI and IoT. We will only be able to realize its potential to the full extent if we ensure a strong focus on customer centricity, go-to-market, monetization, and commercialization.

Taking customer centricity as an example, AI-enabled add-on features can significantly improve the experience of customers who were previously focused purely on physical product performance. AIoT truly has the potential to lift the performance of entire product categories to new levels – both from a UX as well as a quality perspective. The combination of digital and physical will enable us to constantly learn and simulate how products are performing in the field and how customers are using them. This in turn enables us to continuously improve the solution offering – and with it the customer experience. From smart traffic control systems and autonomous driving over drone-based building inspections to ever smarter kitchen appliances, we are already seeing how AIoT is changing the world of physical products. The same applies to the perspective of equipment operators, e.g., manufacturers, rail companies, or energy grid operators, who utilize this new paradigm to constantly optimize their OEE (overall equipment effectiveness).

However, for every real AIoT success, there are many stories of companies that run only prototypes and pilots, but never truly see this scaling to a level where it has real business impact. Creating a sustainable and scalable business enabled by AIoT is not easy. It takes a good combination of business acumen and technical execution capabilities to succeed. Having a strong focus on customer benefits is vital for success. Do not start an AIoT project without clear customer benefit definition which addresses a relevant customer or user pain point. Early in the project focus on commercialization models as well as the go-to-market model.

Digital-enabled business models for physical products require new approaches for sales and marketing. For a startup, this might sound natural, but for an incumbent, it requires a thorough analysis of the existing sales processes and available skills in the sales team. Getting the product–market fit right and crossing the chasm from early adopters to a broader customer base requires a flexible business strategy and the ability to combine digital and physical capabilities. This is not an easy task – neither for a digital start-up nor a manufacturing incumbent – because you often need the culture and capabilities of both types of organizations.

This is where the AIoT User Group and the *Digital Playbook* come in. Bosch has helped to initiate the User Group to bring together good practices from players in different domains and industries. Today, the *Digital Playbook* provides a 360-degree perspective on smart, connected products and solutions:

- How are AI and IoT enabling new business models?
- What are the specific business opportunities for OEMs vs. equipment operators?
- How can these new business models scale up to a level where they become relevant?
- How can matching commercialization and go-to-market strategies be found?
- How can co-creation, sourcing, and legal aspects be managed?
- How to set up a delivery organization that combines data-centric, AI-centric, and software-centric teams
- How can agility, security, functional safety, and robustness be ensured in an integrated DevOps cycle?

The Digital Playbook addresses managers with a digital transformation agenda, product managers, project managers, and solutions architects. It provides teams with a set of good practices, a common understanding, and a common language, to help with the successful application of AIoT and the creation of scalable businesses. A large number of real-world examples and detailed case studies help to ensure a high level of relevance and pragmatism – and make it tangible.

The Digital Playbook would not have been possible without the AIoT User Group and the many experts involved. It is my pleasure to work with such experienced experts in the AIoT User Group's editorial board, which was initiated by my esteemed predecessor Michael Bolle – thank you. Special thanks to Dirk Slama (Bosch and Ferdinand-Steinbeis-Institute), who, as editor-in-chief, is responsible for making *The Digital Playbook* a reality. Many thanks also to the other editors of the print edition: Sebastian Thrun (Udacity), Ulrich Homann (Microsoft), and Heiner Lasi (Ferdinand-Steinbeis-Institute). We also thank the other members of the AIoT editorial board: Prith Banerjee (ANSYS), Jan Bosch (Chalmers University), Ken Forster (Momenta Partners), Dominic Kurtaz (Dassault Systemes), Zara Riahi (Contilio), and Nil Willetts (TM Forum). Finally, my sincere thanks to all the other domain and technology experts who are listed in the AIoT Expert Network. *The Digital Playbook* would not have been possible without your contributions – thank you!

Chief Digital Officer, Robert Bosch GmbH   Tanja Rückert
Stuttgart, Germany

Digital Playbook and the AIoT User Group

In January 2020, a handful of senior IT experts and enthusiasts from different companies met at the Bosch Connectory in Stuttgart to exchange their experiences and views on AI and IoT. AI was at the peak of a new hype, fueled by Alpha Go, advancements in autonomous driving, and, not to forget, Cambridge Analytica. The general feeling was that – with the exception of autonomous driving – AI had not truly arrived in the world of IoT. Every IoT article or presentation in the last 10 years mentioned predictive maintenance, but in reality, many IoT applications were still much more basic. How could AI be better utilized in the world of physical products, manufacturing, and equipment operations? The workshop was organized as an open exchange, with a mixture of presentations and group discussions. After three days, there was so much excitement about the topic and the way collaboration in the group worked that it was decided to make this a regular thing. The result was the formation of the AIoT User Group, a loosely coupled, nonprofit network of AI and IoT practitioners, who work together to exchange experiences and best practices on the application of AI in the IoT. Throughout 2021, local chapters were set up in Singapore (special thanks to CK and Thomas!), Shanghai (Nǐ hǎo, Gene, and Cherry!), and Chicago (hi Fermin and Hans!).

Over time, it became clear that it would make sense to document the collected wisdom in a good practice framework: this is how *The Digital Playbook* started. In fact, it first started as the AIoT Playbook, with a more technical focus. Over time, the business strategy and execution perspectives were added. The result is now a holistic digital playbook, including the technical AIoT Framework.

Content creation for *The Digital Playbook* and AIoT Framework is driven by experts in different domains (see the AIoT Expert Network). The AIoT Editorial Board provides strategic guidance and management support. The basic working modes are so-called Unplugged Sessions, where the real work on the playbook is happening. All the material is developed as open source content (using CC BY 4.0) and is also used as a foundation for different AIoT-related training courses.

How to Get Involved

If you are interested in joining the AIoT User Group, good starting points are the website aiot.rocks, as well as the AIoT User Group on LinkedIn. The main site for *The Digital Playbook* is simply aiotplaybook.org.

Activities of the AIoT User Group involve:

- Unplugged Sessions: These hands-on sessions bring together participants to work on different topics related to AIoT. The results are consolidated and eventually captured in *The Digital Playbook*.
- Training: Participants of the AIoT User Group are organizing AIoT trainings in different regions, utilizing the content from *The Digital Playbook*.
- AIoT Lab: The AIoT Lab is a virtual lab with an increasingly global footprint. Here, experts and practitioners work together to find practical solutions to different AIoT-related problems.

Vision

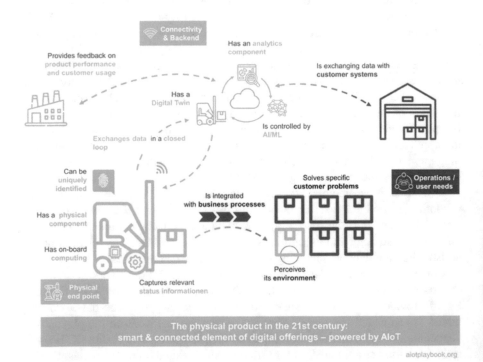

Fig. 1 AIoT Vision

The AIoT User Group has worked on developing a common vision for digital business with smart connected products and solutions, enabled by AIoT (Fig. 1). What does this vision look like? Heiko Löffler and David Monzel are senior consultants at mm1 consulting and frequent contributors to *The Digital Playbook*. Their explanation is as follows:

AIoT combines AI and connectivity for physical products. These can be new product categories (smart, connected products - short SPCs), or retrofit solutions for existing assets and equipment in the field. The general idea is summarized in the figure below: physical products (e.g., a forklift) are end points with physical components and on-board computing (combining hardware, software and AI). The physical component has a unique identifier and continuously captures status and process-critical data, as well as data about its environment. These data are both processed on the device (e.g., via AI) and transmitted to the cloud/backend via IoT connectivity. The product is integrated with business processes (e.g., warehousing), and solves specific customer problems (e.g., optimizing warehousing tasks). The product is exchanging data in a closed loop with a backend (e.g., cloud or on-premises systems). In the cloud, the data are processed to create a Digital Twin of the physical component. This is then continuously analyzed by AI to derive and communicate measures or predictions for both the individual physical components and the entire fleet of products. Furthermore, both the physical and cloud components are integrated with the customer environment to ensure customer-centric value delivery. SCPs become part of the customer's physical and business processes: they sense and interact with their physical environment and are connected to the customer's IT infrastructure (e.g., ERP systems). Finally, the product provides its manufacturer with information about the performance of the product in the field, and how customers are using the product. In all of this, AI can enable new functionality either onboard the product or in the backend. IoT provides the required connectivity between the product and the backend. Digital Twins are a digital representation of the real, physical product — providing abstraction, standardization and a rich, semantic view of the AIoT data. AI can be used to help create Digital Twins, or to build applications that utilize them. Of course, this is a big vision, which will not become a reality for each product category overnight. However, it shows the potential of AIoT. Additionally, not all projects might look at such a high level of productization and deep integration — AIoT can also support more basic retrofit approaches (referred to as solutions throughout the playbook).

About This Book

If you are reading *The Digital Playbook* online or as a PDF, you might sometimes find that not everything is perfect, like in a fully edited book. The reason is that the playbook is constantly evolving, so the decision was made to use a Wiki as the foundation for the playbook. The other book formats are derived as snapshots from the wiki, and the conversion is sometimes not perfect. Additionally, some content might sometimes still not be quite ready. This was a tradeoff between having the perfect book versus the timely delivery of an open, digital content collection that can evolve over time. Since the book formats are published in the open access format, this should hopefully also be acceptable to all readers of the offline versions.

To ensure that you will get the most out of *The Digital Playbook*, we will provide a short overview of the structure of the book, describe its key plays, and provide recommendations on how to best read it.

Structure of *the Digital Playbook*

How can smart, connected products and solutions be enabled with AIoT? *The Digital Playbook* addresses this on two levels (Fig. 2). First, AIoT 101 provides an overview of the relevant core concepts and technologies, including AI, data, digital twins, IoT, and hardware. This helps create a common understanding and common language within a team. Second, *The Digital Playbook* provides a rich set of good practices and templates to help master business strategy, business execution, and technical execution for AIoT-enabled products and solutions. These three areas need to be closely aligned, which is also supported by the playbook.

Heiko Löffler and David Monzel from mm1 have the following take on this:

> In the development and operation of AIoT-enabled systems, many new issues have to be considered, especially for traditional product companies. As a result, new competencies and skills must be acquired, which represents a central challenge. The Digital Playbook gives companies a clear overview of the topics that need to be added to the physical product component to identify individual competence gaps. Underlying these, there are then a

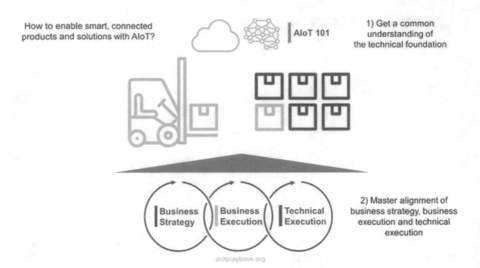

Fig. 2 Why a Digital Playbook?

number of challenges that companies must address to ensure successful implementation. Examples include adapting the business model, building scalable hardware, software and AI architecture, and transforming sales. The Digital Playbook specifically addresses these challenge areas and serves as a comprehensive framework for the realization of smart connected products and solutions.

Key Plays of the Digital Playbook

The Digital Playbook aims to support a holistic and realistic approach for creating and operating AIoT-enabled products and solutions, including:

- AIoT 101: Provides common language and understanding of key concepts, including AI, IoT, digital twin, and data.
- Business Strategy: What is a suitable strategy for AIoT-enabled products and services, addressing the market perspective as well as key internal aspects such as innovation management and target organization?
- Business Execution: From design to procurement and operations – how can an AIoT initiative be executed on the business side?
- Technical Execution – AIoT Framework: From agile to architecture and DevOps – how can an AIoT initiative be executed on the technology side?

Figure 3 provides an overview of all the key plays of *The Digital Playbook*. It is in essence the visual Table of Contents of the playbook. The only sections not shown here are the AIoT case studies, which can be found in the final part of *The Digital Playbook*.

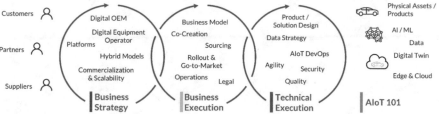

Fig. 3 Overview of *The Digital Playbook*

How to Read This Book

To successfully manage a digital transformation initiative utilizing AIoT, it is important for many stakeholders, such as project managers, product managers and solutions architects, to have a high-level, 360-degree understanding of what is happening in their project or product organization so that they can manage all dependencies and provide a matching structure. *The Digital Playbook* aims to provide a comprehensive, 360-degree overview for exactly this purpose. However, this sometimes means that there is too much content. The playbook is kept as visual as possible, which should enable the reader to browse through the entire playbook by focusing on the diagrams, maybe reading the detailed descriptions only where more details are needed. It can be a good idea to start with a lightweight skimming of the entire structure from A to Z, before then taking more time for the details.

If you are already familiar with the basic concepts of AI, IoT, digital twin, and so on, we recommend still browsing through the 101 chapters: everything is kept very visual, and the different images help make some of the basic assumptions we are making transparent. Also, this section helps with creating a common language.

The business execution and technology execution parts are both kept on a level where they hopefully make sense to both business/domain-focused experts and technology experts. Having a general understanding of both sides seems important for all the aforementioned stakeholders, to ensure that the key perspectives can be closely aligned and that a general understanding of the problems of the other side exists.

For example, in one of the previous AIoT trainings, one of the more technical students asked why he should bother learning about the challenges of sourcing and procurement. It is very easy: first of all, the sourcing team will need input and guidance from the technology experts. Otherwise, the technical team will not obtain the tools and resources it needs. Many projects that neglected those aspects failed due to sourcing-related issues. Conversely, more business-oriented people should also at least skim the technology execution side of the playbook to understand how their counterparts on the technology side are working and what kinds of problems they are facing. *The Digital Playbook* aims to create a level of abstraction that supports this, without becoming too generic.

Enjoy your read, and let us know what you think!

Contents

Part V   Case Studies

Part I
Introduction

The Digital Playbook uses AIoT synonymous with Smart Connected Products and Solutions. Consequently, Part I of the Digital Playbook starts with a discussion of the what, why, how and who of the AIoT. Based on this, we provide an introduction to the core technical topics that are enabling the AIoT. This is structured as a series of "101" chapters, including Artificial Intelligence 101, Data 101, Digital Twin 101, Internet of Things 101, and Hardware 101.

Chapter 1
AIoT 101: What, Why, How, Who

Dirk Slama

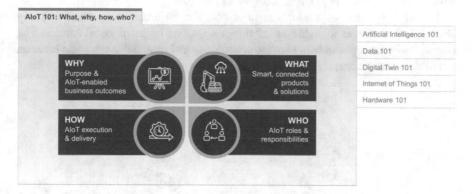

Fig. 1.1 The Why, What, How, and Who of AIoT

AIoT combines two of the most important technology paradigms of the 2020s: Artificial Intelligence (AI) and the Internet of Things (IoT). To best understand AIoT from all relevant perspectives, we will start by looking at the *why, what, who* and *how* perspectives, inspired by the work of Simon Sinek [1] as well as the St. Gallen IoT Lab [2] (Fig. 1.1):

- Why: Better understanding and articulating the purpose and AIoT-enabled business outcomes
- What: What can be achieved with AIoT in terms of smart, connected products and solutions
- Who: Roles and responsibilities in the context of an AIoT initiative

D. Slama (✉)
Ferdinand Steinbeis Institute, Berlin, Germany
e-mail: dirk.slama@bosch.com

- How: Project blueprint for AIoT execution and delivery

While Simon Sinek suggests to *Start with Why*, we will first look at the *what* to provide some context, before discussing *why* you should consider it.

1.1   What: Smart, Connected Products and Solutions with AIoT

The smartness of an AIoT-enabled product or solution is usually related to either an individual physical product/asset ("product/asset intelligence") or to a group/fleet of assets ("swarm intelligence"). Technically, asset intelligence is enabled via edge computing, while swarm intelligence is enabled via cloud computing. Asset intelligence applies AI-algorithms to data that are locally captured and processed (via sensors), while swarm intelligence applies AI-algorithms to data that are captured from multiple assets via IoT technologies in the cloud.

For AIoT systems with a high level of complexity, it can make sense to apply Digital Twin concepts to create a digital representation of the physical entities. The Digital Twin concept can help manage complexity and establish a semantic layer on top of the more technical layers (Fig. 1.2).

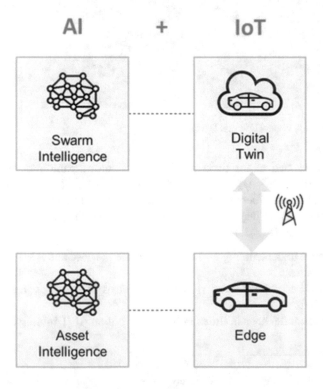

Fig. 1.2 AIoT intro

AI (or Machine Learning, one of the most important subsets of AI) makes use of different types of methods. The three main methods include supervised, unsupervised and reinforcement learning. They are explained in the AIoT 101. Varied, highly specialized AI/ML methods support a wide range of use cases. AIoT focus ones those use cases that are most relevant when dealing with physical products or assets. To mention just one example from the figure shown here, supervised learning can be used for image classification, which plays an important role in optical inspection in manufacturing. While the adoption of AI and ML has already become mainstream in some areas like social media or smartphones, for many AIoT use cases this is still not the case. There is a famous quote from James Bell at Dow Jones, which says that *"Machine Learning is done in Python, AI in PowerPoint."*. The goal of the Digital Playbook is to explore and enable use cases that make use of *real* AI in the context of the IoT, mainly utilizing supervised, unsupervised and reinforcement learning (Fig. 1.3).

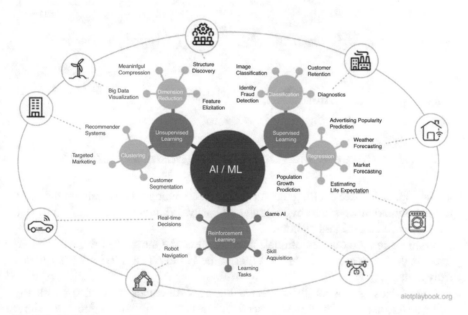

Fig. 1.3 AIoT use case patterns

An important differentiation that we are making in the Digital Playbook is between smart connected products and smart connected solutions. Smart, connected products are often very highly standardized, feature-rich and well rounded. Smart, connected solutions on the other hand are often more custom, ad hoc solutions. They are often designed to solve a specific problem, e.g., for a particular production site, a particular energy grid, etc. Obviously, this is not a black and white differentiation. There are also often cases that are a bit of both product and solution.

As will be discussed in more detail later, smart connected products are manufactured and sold by a Digital OEM, while smart connected solutions are usually acquired and operated by a Digital Equipment Operator. Platforms can also play an important role, even if the platform operator is neither manufacturing nor operating physical assets himself (Fig. 1.4).

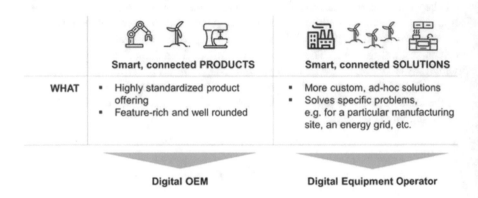

Fig. 1.4 What: product vs. solution

1.2 Why: Purpose and AIoT-Enabled Business Outcomes

While AI and IoT are exciting technical enablers, anybody embarking on the AIoT journey should always start by looking at the *why*: What is the purpose? And what are the expected business outcomes?

From a strategic (and emotional) point of view, the purpose of the AIoT initiative should be clearly articulated: What is the belief? The mission? Why is this truly done?

For business sponsors, the expected business outcomes must also be clearly defined. As discussed in the *what* section, most AIoT initiatives focus on either products or solutions. Depending on the nature of your initiative, the KPIs will differ: AIoT-enabled products tend to focus more on the customer acceptance and revenue side, while AIoT-enabled solutions tend to focus more on efficiency and optimization (Fig. 1.5).

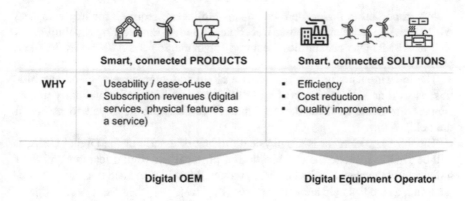

Fig. 1.5 Why: product vs solution

1.3   How: Getting Things (and AI) Done

Smart connected products and solutions usually make use of AI and IoT in different ways. This must be taken into consideration when looking at the *how*. Smart products often rely on AI that was specifically developed for them using a Data Science approach. The goal is often to create new intellectual property that helps differentiate the product. For solutions, this often looks different: here the goal is to minimize development costs, e.g., by reusing existing AI algorithms and models. From the IoT point of view, products and solutions also differ: products usually have built-in connectivity capabilities (line fit), while solutions usually have this capability retrofitted. This is especially important for operators looking at heterogeneous fleets of assets or equipment (Fig. 1.6).

	Smart, connected PRODUCTS	Smart, connected SOLUTIONS
Artificial Intelligence	• Potentially deep Data Science • High potential for new Intellectual Property (IP)	• Ideally high level of re-use, e.g. AI models / libraries
Internet of Things	• Usually line-fit • Often custom IoT hardware	• Usually retro-fit • Usually commercial IoT hardware
	Digital OEM	Digital Equipment Operator

aiotplaybook.org

Fig. 1.6 How: product vs. solution

It is important to understand which capabilities are required for implementing AIoT. The AI side usually requires Data Science and AI Engineering capabilities, as well as AI/ML Ops capabilities (required for managing the AI/ML development process).

The IoT side usually requires generic cloud and edge development capabilities, as well as DevOps supporting both cloud and edge (which usually means support for OTA, or Over-the-Air-Updates of software deployed to assets in the field).

The third key element is the physical product or asset. For the Digital OEM, it will be vital to manage the combination of physical and digital features and their individual life cycles. For the physical product, this will also need to include manufacturing, as well as field support services (Fig. 1.7).

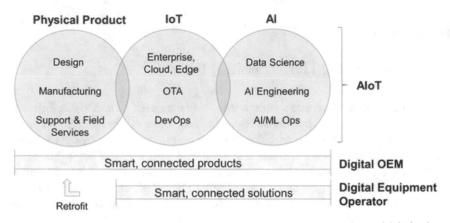

Fig. 1.7 AIoT overview

1.4   Who: AIoT Roles and Responsibilities

The *Who* perspective must address the roles and responsibilities required for successfully delivering your AIoT initiative. These will partially be different for product- vs. solution-centric initiatives, as we will discuss later. It is important to have a holistic view on stakeholder management, including internal and external stakeholders (Fig. 1.8).

Fig. 1.8 AIoT - Who?

External stakeholders can include investors, users of the product of TDB: solution, partner, and suppliers. In a larger organization, internal stakeholders will include business sponsors, senior management, compliance and auditing, legal and tax, global procurement, central IT security, central IT operations, HR, marketing, communication, and sales. Finally, one should not forget about the stakeholders within its own organization, including developers, technology experts, AI experts, and potentially HW/manufacturing (in the case of the Digital OEM).

As indicated in Fig. 1.9, the Digital Playbook primarily addresses middle management, including product/solution managers, project/program managers, development/engineering managers, product/solution architects, security/safety managers, and procurement managers. Ideally, the Playbook should enable these key people to create a common vision and language that enables them to integrate all the other stakeholders.

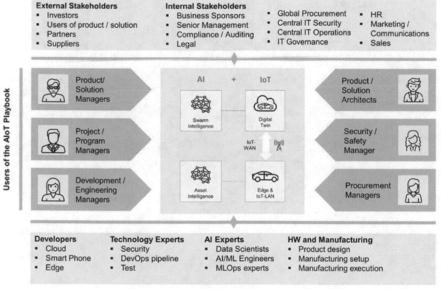

Fig. 1.9 Who: roles and responsibilities

Chapter 2
Artificial Intelligence 101

Dirk Slama

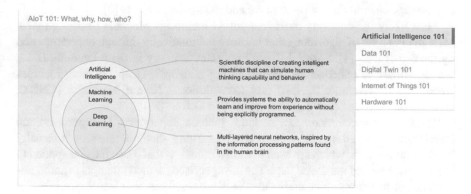

Fig. 2.1 AI 101

This chapter provides an Artificial Intelligence 101, including a basic overview, a summary of Supervised, Unsupervised and Reinforcement Learning, as well as Deep Learning and Artificial Neural Networks (Fig. 2.1).

2.1 Introduction

Artificial Intelligence (AI) is not a new concept. Over the last couple of decades, it has experienced several hype cycles, which alternated with phases of disillusionment and funding cuts ("AI winter"). The massive investments into AI by today's

D. Slama (✉)
Ferdinand Steinbeis Institute, Berlin, Germany
e-mail: dirk.slama@bosch.com

hyperscalers and other companies have significantly fueled the progress made with AI, with many practical applications now being deployed.

A highly visible break-through event was the development of AlphaGo (developed by DeepMind Technologies, which was later acquired by Google), which in 2015 became the first computer Go program to beat a human professional Go player without handicap on a full-sized 19 × 19 Go board. Until then, Go was thought of as being "too deep" for a computer to master on the professional level. AlphaGo uses a combination of machine learning and tree search techniques.

Many modern AI methods are based on advanced statistical methods. However, finding a commonly accepted definition of AI is not easy. A quip in Tesler's Theorem says *"AI is whatever hasn't been done yet"*. As computers are becoming increasingly capable, tasks previously considered to require intelligence are later often removed from the definition of AI. The traditional problems of AI research include reasoning, knowledge representation, planning, learning, natural language processing, perception, and the ability to move and manipulate objects [3].

Most likely the currently most relevant AI method is Machine Learning (ML). ML refers to a set of algorithms that improve automatically through experience and by the use of data [4]. Within ML, an important category is Deep Learning (DL), which utilizes so called *multi-layered neural networks*. Deep Learning includes Deep Neural Networks (DNNs) and Convolutional Neural Networks (CNNs), amongst others. See below for an example of a CNN.

The three most common ML methods include Supervised, Unsupervised and Reinforcement Learning. The Supervised Learning method relies on manually labeled sample data, which are used to train a model so that it can then be applied to similar, but new and unlabeled data. The unsupervised method attempts to automatically detect structures and patterns in data. With reinforcement learning, a trial and error approach is combined with rewards or penalties. Each method is discussed in more detail in the following sections. Some of the key concepts common to these ML methods are summarized in the table following (Fig. 2.2).

Term	Explanation	Example
Instance	The object about which the AI should make a prediction	An image from a vehicle front camera, which needs to be classified as „contains obstacle" or not
Inference	The process of using a trained model to make predictions about an instance based on new data	Apply an ML model to a new image from the vehicle camera to identify potential obstacles
Label	The outcome of a prediction task (either supplied by the training data or by the AI)	Parts of an image labeled as „traffic light", „pedestrian", „speed limit"
Labeling	The process of manually (at least initially) detecting and tagging data samples as input for model training	Manually identify and tag potential obstacles on a large set of images
Training	Machine learning models are trained by using large, representative sets of data, e.g. labeled training data	Manually identify and tag potential obstacles on a large set of images
Feature	An measurable property or characteristic of an instance	Edges and objects in an image
Model	A statistical representation of a prediction task	A model to predict road traffic
Pipeline	The IT infrastructure for an AI/ML algorithm, including data and model management.	The pipline to manage data flows and prediction model definitions

aiotplaybook.org

Fig. 2.2 Key AI terms and definitions

2.2   Supervised Learning

The first AI/ML method we want to look at is Supervised Learning. Supervised Learning requires a data set with some observations (e.g., images) and the labels of the observations (e.g., classes of objects on these images, such as "traffic light", "pedestrian", "speed limit", etc.) (Fig. 2.3).

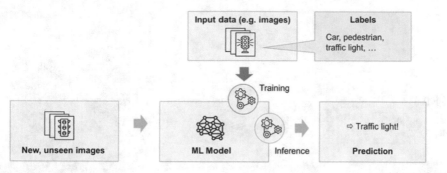

Fig. 2.3 Supervised learning

The models are trained on these labeled data sets, and can then be applied to previously unknown observations. The supervised learning algorithm produces an inference function to make predictions about new, unseen observations that are provided as input. The model can be improved further by comparing its actual output with the intended output: so-called "backward propagation" of errors.

The two main types of supervised models are regression and classification:

- Classification: The output variable is a category, e.g., "stop sign", "traffic light", etc.
- Regression: The output variable is a real continuous value, e.g., electricity demand prediction

Some widely used examples of supervised machine learning algorithms are:

- Linear regression, mainly used for regression problems
- Random forest, mainly used for classification and regression problems
- Support vector machines, mainly used for classification problems

2.3   Unsupervised Learning

The next ML method is Unsupervised Learning, which is a type of algorithm that learns patterns from unlabeled data. The main goal is to uncover previously unknown patterns in data. Unsupervised Machine Learning is used when one has no data on desired outcomes (Fig. 2.4).

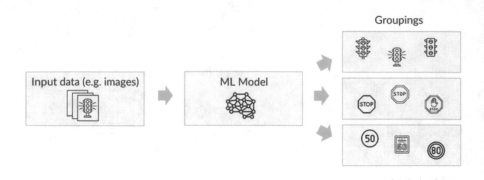

Fig. 2.4 Unsupervised learning

Typical applications of Unsupervised Machine learning include the following:

- Clustering: automatically split the data set into groups according to similarity (not always easy)
- Anomaly detection: used to automatically discover unusual data points in a data set, e.g., to identify a problem with a physical asset or equipment.
- Association mining: used to identify sets of items that frequently occur together in a data set, e.g., "people who buy X also tend to buy Y"
- Latent variable models: commonly used for data preprocessing, e.g., reducing the number of features in a data set (dimensionality reduction)

2.4   Reinforcement Learning

The third common ML method is Reinforcement Learning (RL). In RL, a so-called Agent learns to achieve its goals in an uncertain, potentially complex environment. This can be, for example, a game-like situation, where the agent is deployed into a simulation where it receives rewards or penalties for the actions it performs. The goal of the agent is to maximize the total reward (Fig. 2.5).

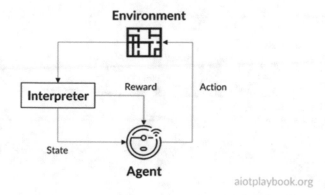

Fig. 2.5   Reinforcement learning

One main challenge in Reinforcement Learning is to create a suitable simulation environment. For example, the RL environment for training autonomous driving algorithms must realistically simulate situations such as braking and collisions. The benefit is that it is usually much cheaper to train the model in a simulated environment, rather than risking damage to real physical objects by using immature models. The challenge is then to transfer the model out of the training environment and into the real world.

2.5   Deep Learning and Artificial Neural Networks

A specialized area within Machine Learning are so-called Artificial Neutral Networks or ANNs (often simply called Neural Networks). ANNs are vaguely inspired by the neural networks that constitute biological brains. An ANN is represented by a collection of connected nodes called neurons. The connections are referred to as edges. Each edge can transmit signals to other neurons (similar to the synapses in the human brain). The receiving neuron processes the incoming signal, and then signals other neurons that are connected to it. Signals are numbers, which are computed by statistical functions.

The relationship between neurons and edges is usually weighted, increasing or decreasing the strength of the signals. The weights can be adjusted as learning proceeds. Usually, neurons are aggregated into layers, where different layers perform different transformations on their input signals. Signals travel through these layers, potentially multiple times. The adjective "deep" in Deep Learning is referring to the use of multiple layers in these networks.

A popular implementation of ANNs are Convolutional Neural Networks (CNNs), which are often used for processing visual and other two-dimensional data. Another example is Generative Adversarial Networks, where multiple networks compete with each other (e.g., in games) (Fig. 2.6).

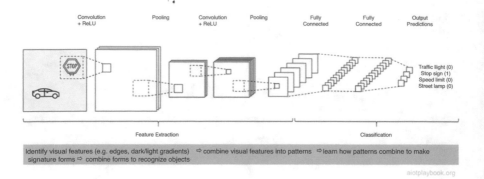

Fig. 2.6 Example: convolutional neural network

The example shows a CNN and its multiple layers. It is used to classify areas of an input image into different categories such as "traffic light" or "stop sign". There are four main operations in this CNN:

- Convolution: Extract features from the input image, preserving the spatial relationship between pixels by using small squares of input data. Convolution is a linear operation: it performs elementwise matrix multiplication and addition.
- Non Linearity: ReLU (Rectified Linear Unit) is an operation applied after the convolution operations. ReLU introduces non-linearity in the CNN, which is important because most real-world data are non-linear.
- Spatial Pooling/down-sampling: This step reduces the dimensionality of each feature map, while retaining the most important information.
- Classification (Fully Connected Layer): The outputs from the first three layers are high-level features of the input image. The Fully Connected Layer uses these features to classify the input image into various classes based on the training dataset.

A more detailed explanation of a similar example is provided by Ujjwal Karn on KDNuggets.

2.6 Summary: AI & Data Analytics

The field of data analytics has evolved over the past decades, and is much broader than just AI and data science - so it is important to understand where AI/ML is fitting in. From the point of view of most AIoT use cases, there are four main types of analytics: descriptive, diagnostic, predictive and prescriptive analytics. Descriptive analytics is the most basic one, using mostly visual analytics to address the question *"What happened?"*. Diagnostic analytics utilizes data mining techniques to answer the question *"Why did it happen?"*, providing some king of root cause analysis. Data mining is the starting point of data science, with its own specific methods, processes, platforms and algorithms. AI - predominantly ML - often addresses the questions *"What is likely to happen?"* and *"What to do about it?"*. Predictive analytics provides forecasts and predictions. Prescriptive analytics can be utilized, for example, to obtain detailed recommendations as work instructions, or even to enable closed-loop automation (Fig. 2.7).

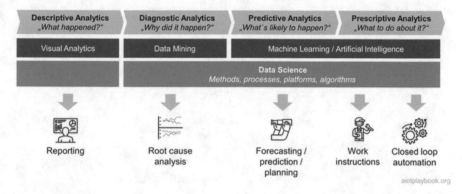

Fig. 2.7 Analytics

Chapter 3
Data 101

Dirk Slama

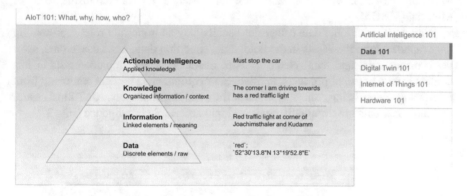

Fig. 3.1 Data 101

Data are the foundation for almost all digital business models. AIoT adds sensor-generated data to the picture. However, in its rawest form, data are usually not usable. Developers, data engineers, analytics experts, and data scientists are working on creating information from data by linking relevant data elements and giving them meaning. By adding context, knowledge is created [5]. In the case of AIoT, knowledge is the foundation of actionable intelligence (Fig. 3.1).

Data are a complex topic with many facets. Data 101 looks at it through different perspectives, including the enterprise perspective, the Data Management, Data Engineering, Data Science, and Domain Knowledge perspectives, and finally the AIoT perspective. Later, the AIoT Data Strategy section will provide an overview of how to implement this in the context of an AIoT initiative.

D. Slama (✉)
Ferdinand Steinbeis Institute, Berlin, Germany
e-mail: dirk.slama@bosch.com

© The Author(s) 2023
D. Slama et al. (eds.), *The Digital Playbook*,
https://doi.org/10.1007/978-3-030-88221-1_3

3.1   Enterprise Data

Traditionally, enterprise data are divided into three main categories: master data, transactional data, and analytics data. Master data are data related to business entities such as customers, products, and financial structures (e.g., cost centers). Master Data Management (MDM) aims to provide a holistic view of all the master data in an enterprise, addressing redundancies and inconsistencies. Transactional data is data related to business events, e.g., the sale of a product or the payment of an invoice. Analytics data are related to business performance, e.g., sales performance of different products in different regions.

From the product perspective, PLM (Product Lifecycle Management) data play an important role. This includes traditionally designed data (including construction models, maintenance instructions, etc.), as well as the generic Engineering Bill of Material (EBOM), and for each product instance a Manufacturing Bill of Material (MBOM).

With AIoT, additional data categories usually play an important role, representing data captured from the assets in the field: asset condition data, asset usage data, asset performance data, and data related to asset maintenance and repair. Assets in this context can be physical products, appliances or equipment. The data can come from interfacing with existing control systems or from additional sensors. AIoT must ensure that these raw data are eventually converted into actionable intelligence (Fig. 3.2).

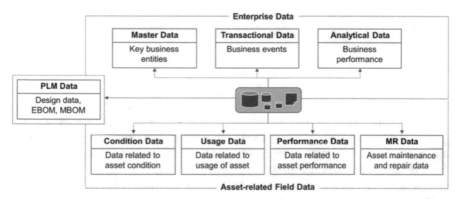

aiotplaybook.org

Fig. 3.2 Data - Enterprise Perspective

3.2   Data Management

Because of the need to efficiently manage large amounts of data, many different databases and other data management systems have been developed. They differ in many ways, including scalability, performance, reliability, and ability to manage data consistency.

For decades, relational database management systems (RDBMS) were the de facto standard. RDBMS manage data in tabular form, i.e., as a collection of tables with each table consisting of a set of rows and columns. They provide many tools and APIs (application programming interfaces) to query, read, create and manipulate data. Most RDBMS support so-called ACID transactions. ACID relates to *Atomicity, Consistency, Isolation, and Durability*. ACID transactions guarantee the validity of data even in the case of fatal errors, e.g., an error during a transfer of funds from one account to another. Most RDBMS support the Structure Query Language (SQL) for queries and updates.

With the emergence of so-called NoSQL databases in the 2010s, the quasi-monopoly of the RDBMS/SQL paradigm ended. While RDBMS are still dominant for transactional data, many projects are now relying on alternative or at least additional databases and data management systems for specific purposes. Examples of NoSQL databases include column databases, key-value databases, graph databases, and document databases.

Column (or wide-column) databases group and store data in columns instead of rows. Since they have neither predefined keys nor column names, they are very flexible and allow for storing large amounts of data within a single column. This allows them to scale easily, even across multiple servers. Document-oriented databases store data in documents, which can also be interlinked. They are very flexible because there is no dedicated schema required for the different documents. Also, they make development very efficient since modern programming languages such as JavaScript provide native support for document formats such as JSON. Key-value databases are very simple but also very scalable. They have a dictionary data structure for storing objects with a unique key. Objects are retrieved only via key lookup. Finally, graph databases store complex graphs of objects, supporting very efficient graph operations. They are most suitable for use cases where many graph operations are required, e.g., in a social network (Fig. 3.3).

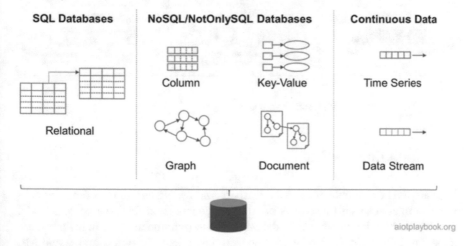

Fig. 3.3 Data - DBMS Perspective

3.3   Analytics Platforms

In addition to the operational systems utilizing the different types of data management systems, analytics was always an important use case. In the 1990s, Data Warehousing systems emerged. They aggregated data from different operational and external systems, and ingested the data via a so-called *"Extract/Transform/Load"* process. The results were data marts, which were optimized for efficient data analytics, using specialized BI (Business Intelligence) and reporting tools. Most Data Warehousing platforms were very much focused on the relational data model.

In the 2010s, Data Lakes emerged. The basic idea was to aggregate all relevant data in one place, including structured (usually relational), non-structured and semi-structured data. Data lakes can be accessed using a number of different tools, including ML/Data Science tools, as well as more traditional BI/reporting tools.

Data lakes were usually designed for batch processing. Many IoT use cases require near real-time processing of streaming and time series data. A number of specialized tools and stream data management platforms have emerged to support this.

From an AIoT point of view, the goal is to eventually merge big data/batch processing with real-time streaming analytics into a single platform to reduce overheads and minimize redundancies (Fig. 3.4).

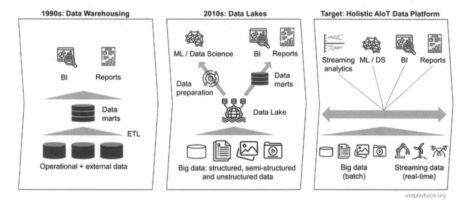

Fig. 3.4 Data Analytics Architecture Evolution

3.4   Data Engineering

Data are the key ingredient for AI. AI expert Andrew Ng has gone as far as launching a campaign to shift the focus of AI practitioners from focusing on ML model development to the quality of the data they use to train the models. In his presentations, he defines the split of work between data-related activities and actual ML model development as 80:20 - this means that 80% of the time and resources are spent on data sourcing and preparation. Building a data pipeline based on a robust and scalable set of data processing tools and platforms is key for success (Fig. 3.5).

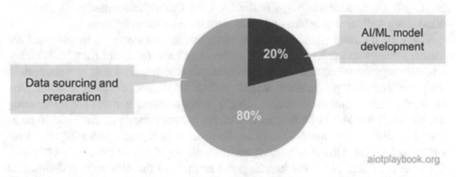

Fig. 3.5 Data vs Model Development

3.4.1   Data Pipeline

From an AIoT point of view, data will play a central role in making products and services 'smart'. In the early stages of the AIoT initiative, the data domain needs to be analysed (see Data Domain Model) to understand the big picture of which data are required/available, and where it resides from a physical/organizational point of view. Depending on the specifics, some aspects of the data domain should also be modeled in more detail to ensure a common understanding. A high-level data architecture should govern how data are collected, stored, integrated, and used. For all data, it must be understood how it can be accessed and secured. A data-centric integration architecture will complete the big picture.

The general setup of the data management for an AIoT initiative will probably differentiate between online and offline use of data. Online relates to data that come from live systems or assets in the field; sometimes also a dedicated test lab. Offline is data (usually data sets) made available to the data engineers and data scientists to create the ML models.

Online work with data will have to follow the usual enterprise rules of data management, including dealing with data storage at scale, data compaction, data retirement, and so on.

The offline work with data (from an ML perspective) usually follows a number of different steps, including data ingestion, data exploration and data preparation. Parallel to all of this, data cataloging, data versioning and lineage, and meta-data management will have to be done.

Data ingestion means the collection of the required data from different sources, including batch data import and data stream ingestion. Typically, this can already include some basic filtering and cleansing. Finally, for data set generation, the data need to be routed to the appropriate data stores.

The ingested data then must be explored. Initial data exploration will focus on the quality of the data and measurements. Data quality can be assessed in several different ways, including frequency counts, descriptive statistics (mean, standard deviation, median), normality (skewness, kurtosis, frequency histograms), etc. Exploratory data analysis helps understand the main characteristics of the data, often using statistical graphics and other data visualization methods.

Based on the findings of the data exploration, the data need to be prepared for further analysis and processing. Data preparation includes data fusion, data cleaning, data augmentation, and finally the creation of the required data sets. Important data cleaning and preparation techniques include basic cleaning ("color" vs. "colour"), entity resolution (determining whether multiple records are referencing the same real-world entity), de-duplication (eliminating redundancies) and imputation. In statistics, imputation describes the process of replacing missing data with substituted values. This is important, because missing data can introduce a substantial amount of bias, make the handling and analysis of the data more arduous, and create reductions in efficiency.

One big caveat regarding data preparation: if the data sets used for AI model training are too much different from the production data against which the models are used later on (inference), there is a danger that the models will not properly work in production. This is why in Fig. 3.6, automated data preparation occurs online before data extraction for data set creation.

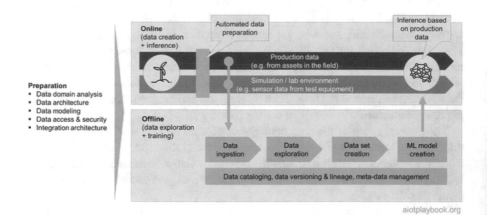

Fig. 3.6 Data - AIoT Perspective

3.4.2   Edge Vs. Cloud

In AIoT, a major concern from the data engineering perspective is the distribution of the data flow and data processing logic between edge and cloud. Sensor-based systems that attempt to apply a cloud-only intelligence strategy need to send all data from all sensors to the cloud for processing and analytics. The advantage of this approach is that no data are lost, and the analytics algorithm can be applied to a full set of data. However, the disadvantages are potentially quite severe: massive consumption of bandwidth, storage capacities and power consumption, as well as high latency (with respect to reacting to the analytics results).

This is why most AIoT designs combine edge intelligence with cloud intelligence. On the edge, the sensor data are pre-processed and filtered. This can result in triggers and alerts, e.g., if thresholds are exceeded or critical patterns in the data stream are detected. Local decisions can be made, allowing us to react in near-real time, which is important in critical situations, or where UX is key. Based on the learnings from the edge intelligence, the edge nodes can make selected data available to the cloud. This can include semantically rich events (e.g., an interpretation of the sensor data), as well as selected rich sample data for further processing in the cloud. In the cloud, more advanced analysis (e.g., predictive or prescriptive) can be applied, taking additional context data into consideration.

The benefits are clear: significant reduction in bandwidth, storage capacities and power consumption, plus faster response times. The intelligent edge cloud continuum takes traditional signal chains to a higher level. However, the basic analog signal chain circuit design philosophy should still be taken into consideration. In addition, the combination of cloud/edge and distributed system engineering expertise with a deep domain and application expertise must be ensured for success (Fig. 3.7).

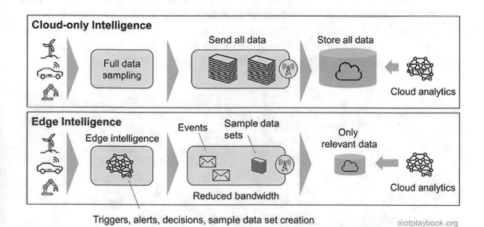

Fig. 3.7 Edge Intelligence

In Fig. 3.8, an intelligent sensor node is monitoring machine vibration. A threshold has been defined. If this threshold is exceeded, a trigger event will notify the backend, including sample data, to provide more insights into the current situation. This data will allow to analyze the status quo. An important question is: will this be sufficient for root cause analysis? Most likely, the system will also have to store vibration data for a given period of time so that in the event of a threshold breach, some data preceding the event can be provided as well, enabling root cause analysis.

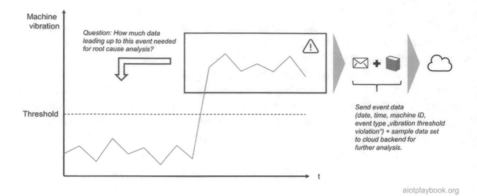

Fig. 3.8 Threshold event and sample data

3.4.3 The Big Loop

For some AIoT systems, it can be quite challenging to capture data representing all possible situations that need to be addressed by the system. This is especially true if the system must deal with very complex and frequently changing environments, and aims to have a high level of accuracy or automation. This is true, for example, for automated driving.

In order to deal with the many different and potentially difficult situations such a system has to handle some companies are implementing what is sometimes called "the big loop": a loop which can constantly capture new, relevant scenarios that the system is not yet able to handle, feed these new scenarios into the machine learning algorithms for retraining, and update the assets in the field with the new models.

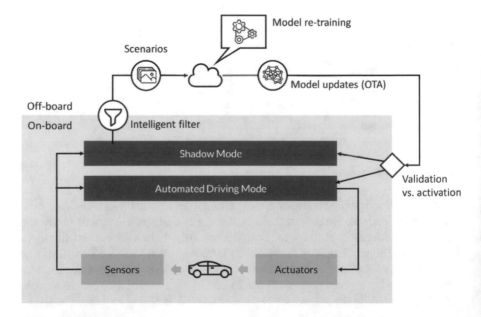

Fig. 3.9 The Big Loop

Figure 3.9 describes how this can be done for automated driving: the system has an Automated Driving Mode, which receives input from different sensors, e.g., cameras, radar, lidar and microphones. This input is processed via sensor data fusion and eventually fed to the AI, which uses the data to plan the vehicle's trajectory. Based on the calculated trajectory, the actuators of the vehicle are instructed, e.g., steering, accelerating and braking. So far so good. In addition, the system has a so-called Shadow Mode. This Shadow Mode is doing pretty much the same calculations as the Automated Driving Mode, except that it does not actually control the vehicle. However, the Shadow Mode is smart in that it recognizes situations that can either not be handled by the AI or where the result is deemed to be suboptimal; for example, another vehicle is detected too late, leading to a sharp braking process. In this case, the Shadow Mode can capture the related data as a scenario, which it then feeds back to the training system in the cloud. The cloud collects new scenarios representing new, relevant traffic situations and uses these scenario data to retrain the AI. The retrained models can then be sent back to the vehicles in the field. Initially, these new models can also be run in the Shadow Mode to understand how they are performing in the field without actually having a potentially negative impact on actual drivers since the Shadow Mode does not interfere with the actual driving process. However, the Shadow Mode can provide valuable feedback about the new model instance and can help validate their effectiveness. Once this has been assured, the models can be activated and used in the real Automated Driving Mode.

Since such an approach with potentially millions of vehicles in the field can help deal with massive amounts of sensor data and make these data manageable by filtering out only the relevant scenarios, it is also referred to as Big Loop.

3.5   Data Science

Data scientists need clean data to build and train predictive models. Of course, ML data can take many different forms, including text (e.g., for auto-correction), audio (e.g., for natural language processing), images (e.g., for optical inspection), video (e.g., for security surveillance), time series data (e.g., electricity metering), event series data (e.g., machine events) and even spatiotemporal data (describing a phenomenon in a particular location and period of time, e.g., for traffic predictions). Many ML use cases require that the raw data be labeled. Labels can provide additional context information for the ML algorithm, e.g., labeling of images (image classification).

The following provides a discussion of AIoT data categories, followed by details on how to derive data sets and label the training data.

3.5.1   Understanding AIoT Data Categories and Matching AI Methods

Understanding the basic AIoT Data Categories and their matching AI Methods is key to AIoT project success. *The Digital Playbook* defines five main categories, including snapshot data (e.g., from cameras), event series data (e.g., events from industrial assets), basic time series data (e.g., from a single sensor with one dimension), panel data (time series with multiple dimensions from different basic sensors), and complex panel data (time series with multiple dimensions from different, high-resolution sensors) (Fig. 3.10).

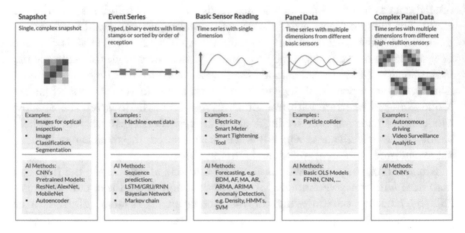

Fig. 3.10   AIoT Data Categories

Figure 3.9 maps some common AI methods to these different AIoT data types, including AF - Autocorrelation Functions, AR – Autoregressive Model, ARIMA – ARMA without stationary condition, ARMA – Mixed Autoregressive Mixed Autoregressive –Moving Average Models, BDM - Basic Deterministic Models, CNN – Convolutional Neural Network, FFNN – Feedforward Neural Network, GRU – Gated recurrent unit, HMM – Hidden Markov Models, LSTM – Long short-term memory, MA – Moving Average, OLS – Ordinary Least Squares, RNN – Recurrent Neural Network, SVM – Support Vector Machine.

3.5.2   Data Sets

In ML projects, we need data sets to train and test the model. A data set is a collection of data, e.g., a set of files or a specific table in a database. For the latter, the rows in the table correspond to members of the data set, while every column of the table represents a particular variable.

The data set is usually split into training (approx. 60%), validation (approx. 20%), and test data sets (approx. 20%). Validation sets are used to select and tune the final ML model by estimating the skill of the tuned model for comparison with other models. Finally, the training data set is used to train a model. The test data set is used to evaluate how well the model was trained (Fig. 3.11).

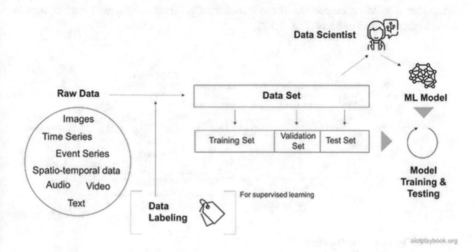

Fig. 3.11 Data - ML Perspective

In the article "From model-centric to data-centric" [6], Fabiana Clemente provides the following guiding questions regarding data preparation:

- Is the data complete?
- Is the data relevant for the use case?
- If labels are available, are they consistent?
- Is the presence of bias impacting the performance?
- Do I have enough data?

In order to succeed in the adoption of a data-centric approach to ML, focusing on these questions will be key.

3.5.3 Data Labeling

Data labeling is required for supervised learning. It usually means that human data labelers manually review training data sets, tagging relevant data with specific labels. For example, this can mean manually reviewing pictures and tagging objects in them, such as cars, people, and traffic signs. A data labeling platform can help to support and streamline the process.

Is data labeling the job of a data scientist? Most likely, not directly. However, the data scientist has to be involved to ensure that the process is set up properly, including the relevant QA processes to avoid bad label data quality or labeled data with a strong bias. Depending on the task at hand, data labeling can be done in-house, outhouse, or by crowdsourcing. This will heavily depend on the data volumes as well as the required skill set. For example, correct labeling of data related to medical diagnostics, building inspection or manufacturing product quality will require input from highly skilled experts (Fig. 3.12).

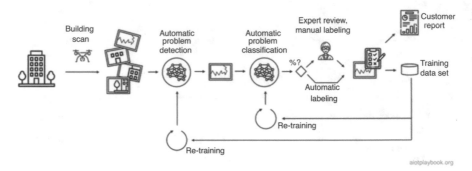

Fig. 3.12 Data Labeling Example

Take, for example, building inspection using data generated from drone-based building scans. This is actually described in detail in the TÜV SÜD building façade inspection case study. Indicators detected in such an application can vary widely, depending on the many different materials and components used for building façades. Large building inspection companies such as TÜV SÜD have many experts for the different combinations of materials and failure categories. Building up a training data set with labeled data for automatically detecting all possible defects requires considerable resources. Such projects typically implement a hybrid solution that combines AI-based automation where there are sufficient training data and manual labeling where there is not. The system will first attempt to automatically detect defects, allowing false positives and minimizing false negatives. The data is then submitted for manual verification. Depending on the expert's opinion the result is accepted or replaced with manual input. The results of this process are then used to further enhance the training dataset and create the problem report for the customer. This example shows a type of labeling process that will require close collaboration between data engineers, data scientists and domain experts.

3.6 Domain Knowledge

One of the biggest challenges in many AI/ML projects is access to the required domain knowledge. Domain knowledge is usually a combination of general business acumen, industry vertical knowledge, and an understanding of the data lineage.

Domain knowledge is essential for creating the right hypotheses that data science can then either prove or disprove. It is also important for interpreting the results of the analyses and modeling work.

One of the most challenging parts of machine learning is feature engineering. Understanding domain-specific variables and how they relate to particular outcomes is key for this. Without a certain level of domain knowledge, it will be difficult to direct the data exploration and support the feature engineering process. Even after the features are generated, it is important to understand the relationships between different variables to effectively perform plausibility checks. Being able to look at the outcome of a model to determine if the result makes sense will be difficult without domain knowledge, which will make quality assurance very difficult.

There have been many discussions about how much domain knowledge the data scientist itself needs, and how much can come from domain experts in the field. The general consensus seems to be that a certain amount of domain knowledge by the data scientist is required and that a team effort where generalist data scientists work together with experienced domain experts usually also works well. This will also heavily depend on the industry. An internet start-up that is all about "clicks" and related concepts will make it easy for data scientists to build domain knowledge. In other industries, such as finance, healthcare or manufacturing, this can be more difficult.

The case study AIoT in High-Volume Manufacturing Network describes how an organization is set up which always aims to team up data science experts with domain experts in factories (referred to as *"tandem teams"*). Another trend here is *"Citizen Data Science"*, which aims to make it easy to use data science tools available directly to domain experts.

In many projects, close alignment between the data science experts and the domain experts is also a prerequisite for trust in the project outcomes. Given that it is often difficult in data science to make the results *"explainable"*, this level of trust is key.

3.7 Chicken Vs. Egg

Finally, a key question for AIoT initiatives is: what comes first, the data or the use case? In theory, any kind of data can be acquired via additional sensors to best support a given use case. In practice, the ability to add more sensors or other data sources is limited due to cost and other considerations. Usually, only greenfield, short tail AIoT initiatives will have the luxury of defining which data to use specifically for their use case. Most long tail AIoT initiatives will have to implement use cases based on already existing data.

For example, the building inspection use case from earlier is a potential short tail opportunity, which will allow the system designers to specify exactly which sensors to deploy on the drone used for the building scans, derived from the use cases which need to be supported. This type of luxury will not be available in many long tail use cases, e.g., in manufacturing optimization as outlined in AIoT and high volume manufacturing case study (Fig. 3.13).

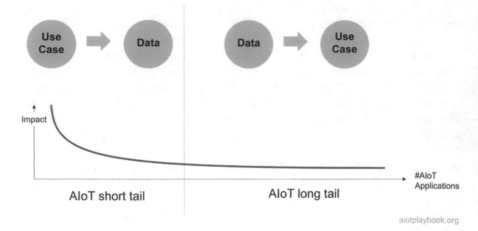

Fig. 3.13 Data - ML Long Tail

Chapter 4
Digital Twin 101

Dirk Slama

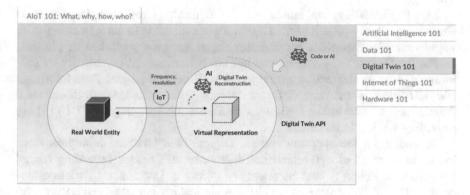

Fig. 4.1 Digital Twin 101

Digital Twins are a digital representation of real-world physical entities. They help manage complexity and establish a semantic layer on top of the more technical layers. This in turn can make it easier to realize business goals and implement AI/ML solutions using machine data. This chapter provides an overview, some concrete examples, as well as a discussion in which situations the Digital Twin approach should be considered for an AIoT initiative (Fig. 4.1).

D. Slama (✉)
Ferdinand Steinbeis Institute, Berlin, Germany
e-mail: dirk.slama@bosch.com

© The Author(s) 2023
D. Slama et al. (eds.), *The Digital Playbook*,
https://doi.org/10.1007/978-3-030-88221-1_4

4.1   Introduction

There are multiple "flavors" of digital twins. The Platform Industrie 4.0 places the Asset Administration Shell at the core of its Digital Twin strategy [7]. Many PLM companies include the 3D CAD data as a part of the Digital Twin. Some advanced definitions of Digital Twin also include physics simulation. The Digital Twin Consortium defines the digital twin as follows: *"A digital twin is a virtual representation of real-world entities and processes, synchronized at a specified frequency and fidelity. Digital twin systems transform business by accelerating holistic understanding, optimal decision-making, and effective action. Digital twins use real-time and historical data to represent the past and present and simulate predicted futures. Digital twins are motivated by outcomes, tailored to use cases, powered by integration, built on data, guided by domain knowledge, and implemented in IT/OT systems"* [8].

The Digital Playbook builds on the definition from the Digital Twin Consortium. A key benefit of the Digital Twin concept is to manage complexity via abstraction. Especially for complex, heterogeneous portfolios of physical assets, the Digital Twin concept can help to better manage complexity by providing a layer of abstraction, e.g. through well-defined Digital Twin interfaces and relationships between different Digital Twin instances. Both the I4.0 AdminShell as well as the Digital Twins Definition Language (DTDL) [9] are providing support in this area.

Depending on the approach chosen, Digital Twin interface definitions often extend the concept of well-established component API models by adding Digital Twin specific concepts such as telemetry events and commands. Relationships between Digital Twin instances can differ. A particularly important one is the aggregation relationship, because this will often be the foundation of managing more complex networks of heterogeneous assets.

The goal of many Digital Twin projects is to create semantic models that allow us to better understand the meaning of information. Ontologies are a concept where reusable, industry-specific libraries of Digital Twin models are created and exchanged to support this (Fig. 4.2).

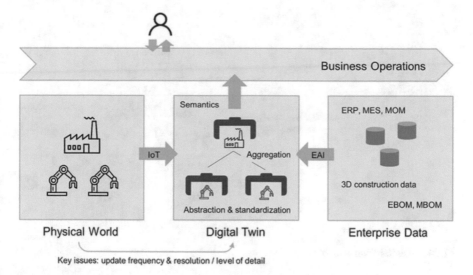

Key issues: update frequency & resolution / level of detail

Fig. 4.2 Digital Twin – overview

4.2 Example

A good example of a Digital Twin is a system that makes route recommendations to drivers of electric vehicles, including stop points at available charging stations. For these recommendations, the system will need a representation of the vehicle itself (including charging status), as well as the charging stations along the chosen route. If this information is logically aggregated as a Digital Twin, the AI in the backend can then use this DT to perform the route calculation, without having to worry about technical integration with the vehicle and the charging stations in the field.

Similarly, the feature responsible for reserving a charging station after a stop has been selected can benefit if the charging station is made available in the form of a Digital Twin, allowing us to make the reservation without having to deal with the underlying complexity of the remote interaction.

The Digital Twin in this case provides a higher level of abstraction than would be made available, for example, via a basic API architecture. This is especially true if the Digital Twin is taking care of data synchronization issues (Fig. 4.3).

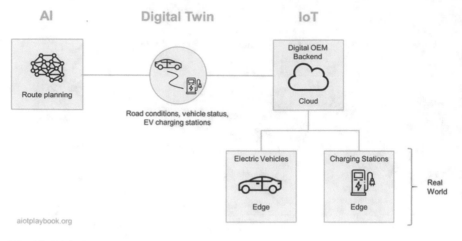

Fig. 4.3 Digital Twin example

4.3   Digital Twin and AIoT

In an AIoT initiative, the Digital Twin concept can play an important role in providing a semantic abstraction layer. The IoT plays the role of providing connectivity services. AI, on the other hand, can play two roles:

- Reconstruction: AI can be an important tool for the reconstruction process; the process of creating (or "reconstructing") the virtual representation based on the raw data from the sensors.
- Application: Once the Digital Twin is reconstructed, another AI algorithm can be applied to the semantically rich representation of the Digital Twin in order to support the business goals (Fig. 4.4)

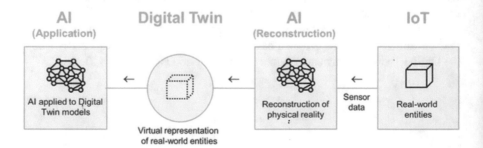

Fig. 4.4 Digital Twin and AIoT

4.3.1 Example 1: Electric Vehicle

The first example to demonstrate this concept is building on the EV scenario from earlier on. In addition, the DT concept is now also applied to the Highly Automated Driving Function of the vehicle, which includes short term trajectory and long-term path planning.

For short-term planning, a digital twin of the vehicle surroundings is created (here, the AI supports the reconstruction of the DT). Next, AI uses the semantically rich interfaces of the digital twin of the vehicle surroundings to perform short-term trajectory planning. This AI will also take the long-term path into consideration, e.g., to determine which way to take on each crossing (Fig. 4.5).

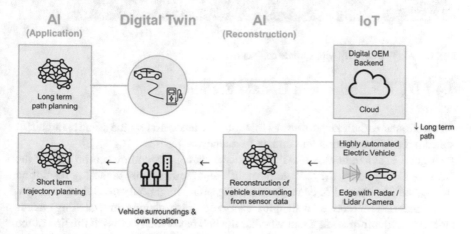

Fig. 4.5 Digital Twin and AIoT – example

4.3.2 Example 2: Particle Collider

The second example is a particle collider, such as the Large Hadron Collider at CERN. The particle collider uses a 3D grid of ruggedized radioactivity sensors in a cavern of the collider to capture radioactivity after the collision. These data are fed into a very complex tier of compute nodes, which are applying advanced analytics concepts to create a digital reconstruction of the particle collision. This Digital Twin is then the foundation of the analysis of the physical phenomena that could be observed (Fig. 4.6).

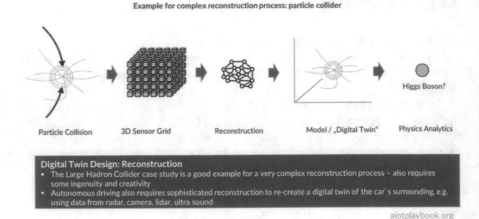

Fig. 4.6 AIoT & Digital Twin: particle collider example

4.4 DT Resolution and Update Frequency

As mentioned earlier, key questions that must be answered by the solution architect concern the DT resolution and update frequency.

A good example here is a DT for a soccer game. Depending on the role of the different stakeholders, they would have different requirements regarding resolution and update frequency. For example, a betting office might only need the final score of the game. The referee (well, plus everybody else playing or watching) needs more detailed information about whether the ball has actually crossed the line of the goal, in case of a shot on the goal. The audience usually wants an even higher "resolution" for the Internet live feed, including all significant events (goals, fouls, etc.). The team coach might require a detailed heat map of the position of each player during every minute of the game. Finally, the team physician wants additional information about the biorhythm of each player during the entire game (Fig. 4.7).

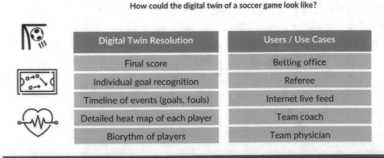

Fig. 4.7 Digital Twin: soccer example

Mark Haberland is the CEO of Clariba. The company is offering custom tracking and analytics solutions for soccer teams. He shares the following insights with us: *Football clubs around the world are striving to achieve a competitive advantage and increase performance using the immense data available from the use of digital technology in every aspect of the sport. Continued innovation in the application of sensors, smart video analytics with edge computing, drones, and even robotics, streaming data via mesh WIFI networks and 5G connectivity is providing incredible new capabilities and opportunities for real-time insights. Harnessing this ever-increasing amount of data will allow forward-looking clubs to experiment and to innovate and develop new algorithms to achieve the insights needed to increase player and team performance to win on the pitch. Investing in new technologies, building in-house capabilities and co-innovating with partners such as universities and specialized companies savvy in AIoT will be a differentiating factor for football organisations that want to lead the way* (Fig. 4.8).

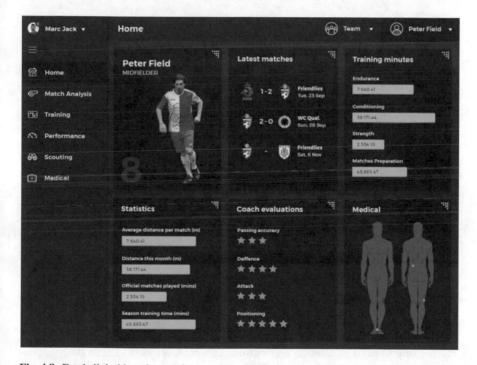

Fig. 4.8 Football dashboard example

So there are a number of key questions here: How high should the resolution of the DT be? And how can a combination of sensors and reconstruction algorithms deliver at this resolution?

For individual goal recognition, a dedicated sensor could be embedded in the ball, with a counterpart in the goal posts. This would require a modification of the ball and the goal posts but would allow for very straightforward reconstruction, e.g., via a simple rule.

Things become more complicated for the reconstruction process if a video camera is used instead. Here, AI/ML could be utilized, e.g., for goal recognition.

For the biorhythm, chances are that a specialized type of sensor will be somehow attached to the player's body, e.g., in his shorts or t-shirt. For the reconstruction process, advanced analytics will probably be required (Fig. 4.9).

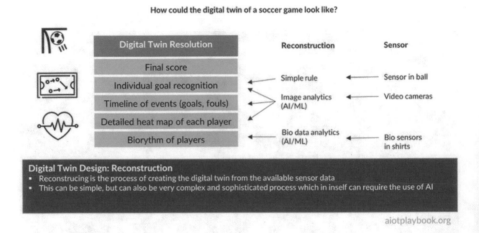

Fig. 4.9 Digital Twin: soccer example (details)

4.5   Advanced Digital Twins: Physics Simulation and Virtual Sensors

To wrap up the introduction of Digital Twins, we will now examine advanced Digital Twins using physics simulation and virtual sensors. A real-world case study is provided, which is summarized in the figure following. Two experts have been interviewed for this: Dr. Przemyslaw Gromala (Team leader of Modelling and Simulation at Bosch Engineering Battery Management Systems and New Products) and Dr. Prith Banerjee (Chief Technology Officer of ANSYS, a provider of multi-physics engineering simulation software for product design, testing and operation) (Fig. 4.10).

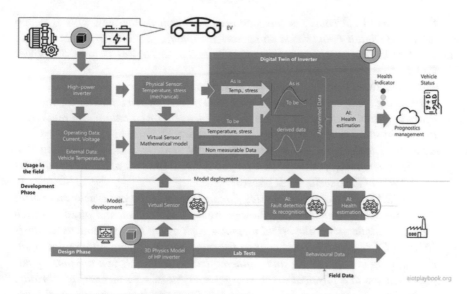

Fig. 4.10 Physics simulation and virtual sensors

Dirk Slama: *Prith, can you share your definition of Digital Twin with us?*

Prith Banerjee: *We are looking at Digital Twins during the design, manufacturing and operations phase. Let's take a car motor as an example. You start by creating a CAD model during the design phase. Using the ANSYS products, you can then also create a physics simulation model, which helps you get insights into the future performance of the motor. Then, after the motor has been manufactured, you are entering the operations phase. During operations, sensors are used on the asset to measure key operational indicators, e.g., vibration and temperature of the motor. You can then calibrate the as-designed value of the asset to match the as-manufactured and as-operated values. In some cases, you might not be able to put real sensors in all the places where you need them, either because it is too costly, or technically not feasible. In this case, you can derive a virtual sensor from the physics simulation model. Combining real sensors with virtual sensors can get you a Digital Twin which represents the real world with a very high level of accuracy.*

Dirk: *Thanks. Przemyslaw, you are working on applying these concepts to high power inverters for electric vehicles. Can you start by explaining what a high-power inverter is, and what use cases you are seeing for advanced Digital Twins?*

Przemyslaw Gromala: *An inverter for electric vehicles is a power electronics system that converts the direct current from the HV-battery into the (3 Phase) alternating current controlling the motor. This of course is a very important component of any EV or hybrid vehicle. One of the major challenges that we are facing right now is that these power modules are completely new devices in automotive electronics, which are combinations of relatively new materials, e.g., epoxy-based molding compounds in combination with silicon carbide technologies, as well as new interconnection technologies based on silver sintering. Of course, we need to understand the complete interactions in a much better way. This is where advanced Digital Twins based on numerical simulations come in hand.*

Dirk: *What kind of sensors are you using to build the Digital Twin of the power inverter?*

Przemyslaw: *There are two categories of sensors. The first category includes real sensors, e.g., temperature sensors, vibration or motion sensors. The second category is virtual sensors, which allow us to measure or calculate different stress and strain states in the locations when real sensors cannot be applied.*

Dirk: *Prith, I understand this concept of virtual sensors is also something that ANSYS is very much focusing on. Can you tell us a little bit about where this fits in, in terms of the development phases?*

Prith: *At ANSYS we do detailed physics simulation. The different sensor categories that Przemyslaw is talking about are going back to fundamental physics, which we are simulating with numerical methods, such as the Finite Elements Analysis (FEA). FEA is a widely used method for numerically solving differential equations arising in engineering and mathematical modeling. If you are putting different sensors to a physical asset, you get many data, e.g., measuring vibration or pressure. The problem is that there are locations inside the physical asset where you cannot place a physical sensor. Take our example, the inverter. If you would put sensors to all the places where you want them, it would become technically impossible and prohibitive from a cost perspective. Therefore, in places where we cannot afford to place a real sensor for technical or cost reasons, we can logically assign a virtual sensor. In addition, this is where simulation comes in. We model the actual physics of the system using Finite Element Analysis to predict how a product reacts to real-world forces, vibration, heat, fluid flow, and other physical effects. The physics-based simulation of virtual sensors allows interpolation and extrapolation of the different values. In our example of a power inverter, we will put physical sensors where we can actually put the sensor. And then we will add the virtual sensors and we will essentially say, "Hey, if I were to have a sensor here, this is what the sensor would have produced." Now, this virtual sensor needs to do the simulation based on some boundary conditions. So what we do is we look at the actual operating data. We look at the current and voltage of the inverter, the actual temperature, and the actual vibration of the car. We will take all those sorts of inputs as boundary conditions for our simulation.*

Now, the next question is how detailed does the simulation have to be? If we would take a full 3D model as the foundation of the physics-based Digital Twins, this could be too much detail to master with realistic effort. This is why we are applying what is called Reduced Order Modeling (ROM). ROM is a very efficient method for reducing the computational complexity of mathematical models in numerical simulations. We can use ROM to design the virtual sensors, and to derive the values we will get from them in different situations. Finally, in the production system, we combine the outputs of the real sensors with the outputs from the virtual sensors by applying Machine Learning algorithms. And this is what gives us a highly accurate Digital Twin using real and virtual sensors together.

Dirk: *Przemyslaw, how are you applying all this to your power inverter?*

Przemyslaw: *What is crucial here are the nonlinear simulations, especially the mechanical simulations. This is because the material behaves very differently, e.g., depending on the temperature and time. For the power inverter, we cannot apply physical sensors at all the places where we would like to have them. This is where the combination of nonlinear simulations, virtual sensors, and machine learning is giving us a real edge. This is the foundation for new applications such as the estimation of the state of health of the devices, including prognostics and health management for power electronics. Once we have this established, we can then even think about predictive maintenance for electronics systems.*

Prith: *Let me add to this. The typical way that people build Digital Twins is by attaching some sort of sensors to an asset, collect a lot of data, and then build an AI model based on that data. What we have found is that the accuracy of the purely ML-based analytics of a Digital Twin is approximately 80%. If you are doing a physics-based simulation of the Digital Twin, you can increase the accuracy to approximately 90%. Now, by combining the ML-based analytics with the physics-based approach into a hybrid Digital Twin – as Przemyslaw has described it in the work we are doing today with Bosch – you can actually increase the accuracy to up to 99%. This means that if you replace an asset worth a hundred thousand dollars based on a prediction that is only 80% accurate, this means you are likely to lose $20,000 on average. And that is a big business cost. If you can reduce that cost, that error to 1%, essentially this waste of $20,000 becomes reduced to only $1000. That is the business value that we are producing with hybrid Digital Twins leveraging AI and physics-based simulation.*

Dirk: *The Digital Twin really supports the entire product life-cycle?*

Przemyslaw: *Yes, this is important. Digital Twins start with the design of our power devices. Then we track what happens during production. Finally, we start with the reliability assessment of our devices in the field until the component actually goes out of the field. And this is the moment when the Digital Twin will reach its end of use. In addition, that means Digital Twin for me, it is right from the beginning of the design process until the end of life of the device.*

Dirk: *During operations, where does the Digital Twin actually reside in your architecture?*

Przemyslaw: *That is a very important point. Especially for applications where the Digital Twin is applied to assets in the field, it is important to have them run on-board the asset, e.g., the car. This means that we are not relying on very large clusters for Digital Twin processing but can use the microprocessor or microcontroller that is running in the car. By implementing the Digital Twins in the car, we do not have to transfer all the data to remote cloud services.*

Prith: *When you build a very high-fidelity version of a Digital Twin, this would usually require a high-performance compute node, e.g., in the cloud. However, what we do is to build a simplified model and then export it. And it is this simplified twin that runs in a run time using docker containers on the edge with a very small memory footprint, as Przemyslaw said. Running it on the edge makes it possible for the twins to operate at the frequency of the real assets.*

Dirk: *Prith, in which industries do you see this being applied, predominantly?*

Prith: *We are currently focused on three broad use cases. One use case we just talked about is for electric vehicles. Another very big use case is for industrial flow networks in oil and gas, where we create a Digital Twin of an oil and gas network with lots of valves and compressors, etc. Another area is manufacturing, e.g., for large injection molding systems and other types of equipment in a factory. So there are lots and lots of applications of Digital Twins across different verticals.*
Dirk: *Thank you!*

Chapter 5
Internet of Things 101

Dirk Slama

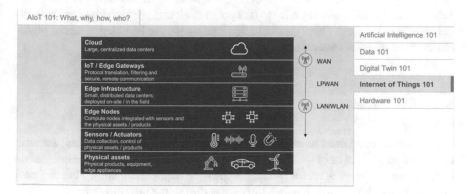

Fig. 5.1 IoT 101

This chapter provides an Internet of Things 101, including a brief overview, a discussion of IoT Sensors and Actuators, IoT Architecture, IoT Protocol Layers, and IoT Connectivity (Fig. 5.1).

D. Slama (✉)
Ferdinand Steinbeis Institute, Berlin, Germany
e-mail: dirk.slama@bosch.com

© The Author(s) 2023
D. Slama et al. (eds.), *The Digital Playbook*,
https://doi.org/10.1007/978-3-030-88221-1_5

5.1   Introduction

The Internet of Things (IoT) describes the concept of connecting physical objects (a.k.a. the "Things"), which are embedded with sensors and actuators over the Internet. This connection can either be direct between physical objects or between physical objects and a back-end data center (cloud or on-premises). Surprisingly, this connectivity will often make use of Internet protocols (IP, UDP, etc.) but use protected enterprise networks instead of the open Internet. The IoT has emerged as a concept following earlier approaches such as M2M (Machine-to-Machine communication) and Telematics, usually adding a richer set of digital services. Industrial Internet of Things (IIoT) refers to industrial IoT applications. Edge computing is quickly emerging as field-based computing capacity closer to the physical objects and as a counterpart to centralized cloud computing. The boundaries between Edge computing, IoT, and embedded computing can sometimes be blurry. The term *IoT Edge Node* is usually used to refer to compute nodes that are embedded with physical objects in the field. Other types of edge equipment can be independent of physical assets, e.g., local data centers.

5.2   IoT Architecture

An IoT architecture must support the creation of a bridge between physical assets in the field and a cloud or on-premises backend. This first challenge is the integration of the actual physical asset (or product, appliance, equipment, etc.). This can usually be done either as part of a line-fit process during manufacturing or as a retrofit process (especially for legacy assets). Asset integration addresses issues such as power supply, ruggedization, antenna positioning, etc.

On – or close to – the asset, the IoT architecture usually positions an edge layer. The first elements here are sensors or actuators (some might argue that this is part of the asset, not part of the edge, which can also be a valid design). In addition, we often find edge applications (e.g., preprocessing of sensor data) and edge AI/Asset Intelligence (e.g., sensor data fusion and autonomous control). Modern edge platforms provide a runtime for local compute resources, as well as a gateway functionality for remote communication.

In the backend, AIoT backend applications and backend AI/swarm intelligence operate on the data received from assets in the field. This is often supported by IoT/IIoT-specific middleware, provided as Platform-as-a-Service (PaaS). Normal Cloud-PaaS services as well as Infrastructure-as-a-Service (IaaS) are required as the foundation.

Not to be underestimated is the need to integrate with existing Enterprise Applications: most IoT projects also have an EAI element (Enterprise Application Integration) (Fig. 5.2).

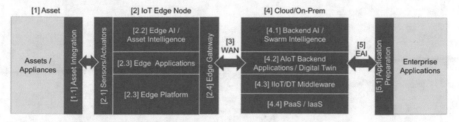

Fig. 5.2 IoT architecture

5.3   IoT Sensors and Actuators

Most IoT systems benefit from the use of sensors and actuators to acquire data from the field and control the behavior of assets in the field. Typical sensor categories include image and video, acoustics and noise, temperature, moisture, light, presence and proximity, motion, gyroscope (rotation and velocity), water level, and chemicals. Typical actuators include motors (servo, stepper, DC), linear actuators, relays (electric switches), and solenoids (electromagnets) (Fig. 5.3).

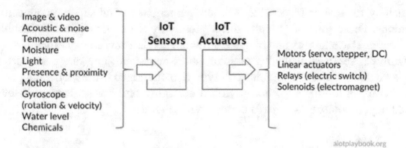

Fig. 5.3 IoT sensors and actuators

5.4   IoT Protocol Layers

Connectivity between the IoT edge nodes and the backend requires different protocol layers. Similar to the postal services with letters and parcels, these protocol layers are responsible for packaging, addressing and routing data in various forms. At the top layers, application-specific protocols are responsible for supporting information exchange at the business level. Lower-level transport protocol layers are responsible for the transfer of anonymous data packages (business-level data are often split into smaller packages for transportation purposes, and then reassembled on the receiver side). The most well-known protocol is TCP/IP, which is the standard protocol on the Internet. IoT communication networks must establish a physical data link between the different nodes involved. This happens using a variety of standardized and proprietary protocols, especially for wireless communication (Fig. 5.4).

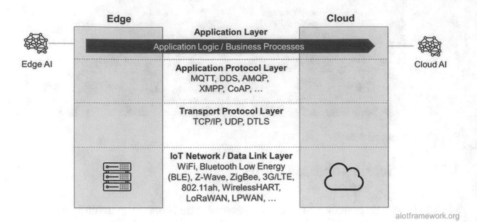

Fig. 5.4 IoT protocol layers

5.5 IoT Connectivity

Especially for mobile physical assets, it is important to find suitable connectivity solutions. These solutions usually differ greatly by a number of factors, including cost, regional availability and range, bandwidth and latency, energy efficiency.

Different services, including cellular, short-range and long-range services are available and can also be combined. A typical pattern is to use a short-range protocol to connect mobile edge nodes with a central (mobile or stationary) gateway, which then establishes wide-area connectivity (Fig. 5.5).

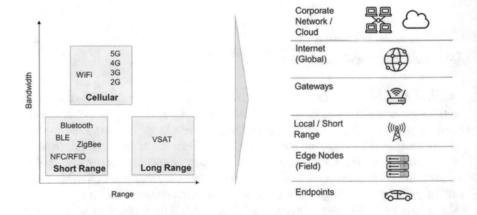

Fig. 5.5 AIoT network architecture

5.6 Over-the-Air Updates

Another key capability of modern IoT systems is to perform updates Over-the-Air (OTA). OTA allows us to update Software (SOTA) or Firmware (FOTA) on a remote asset, e.g., via WLAN or mobile networks. OTA update mechanisms are now a common feature for almost all smartphones, tablets, and similar devices. In automotive systems, some early adopters, such as Tesla, have been pioneering OTA. Currently, most other OEMs have started to adopt OTA updates as well.

OTA Updates are a key capability of any AIoT product since they allow software deployed on the asset to evolve over time, gradually rolling out new functionality. If combined with an app store, OTA updates allow tapping into a rich developer community, which can add new functionalities and apps in many creative ways that sometimes have not been foreseen by the asset manufacturer and platform operator.

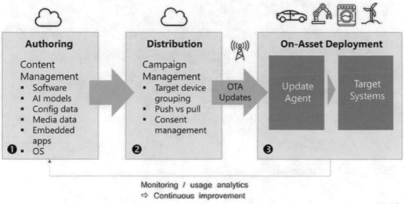

Fig. 5.6 OTA updates – overview

Figure 5.6 shows a typical OTA architecture for AIoT products. It all starts with the authoring (1) of new versions of software or firmware. This process can include deliveries from suppliers and sub-suppliers, which need to be integrated, tested and bundled. Next, the distribution component (2) is responsible for making the updates available in different regions, and coordinating the update campaigns. Finally, on-asset deployment (3) is responsible for ensuring that the update is reaching its target.

Because OTA is becoming such an important feature of IoT and AIoT systems, a number of standards are emerging in this space. ISO 24089 aims to provide a standard architecture for OTA updates for road vehicles. The OMA DM protocol provides an integrated and extensible framework for OTA management. A number of existing standards specifically describe how to implement delta updates for firmware, thus dramatically reducing the amount of data to be transferred to each device individually.

5.6.1 Distribution

The distribution component of an OTA platform is typically responsible for package preparation and campaign management, as well as tracking and reporting. Campaign management can rely on complex rules to control updates to large amounts of devices or assets. For example, rules can help ensure that assets such as road vehicles are not updated in critical situations (e.g., prevent updates while driving, or require the asset to not be in a remote area in case of update problems) (Fig. 5.7).

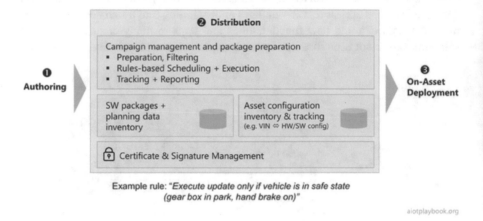

Fig. 5.7 OTA updates – distribution

The typical components of the distribution tier include:

- Repository for firmware/software packages and planning data
- Asset inventory and update tracking/reporting
- Certificate & Signature Management

5.6.2 Deployment

Most OTA platforms include a dedicated update agent, which receives software/ firmware updates from the remote distribution tier in the backend. Key functionality of the update agent typically include Download Management, Security Management (including management and validation of security certificates and signatures), and distribution of the incoming updates to the target compute nodes (Fig. 5.8).

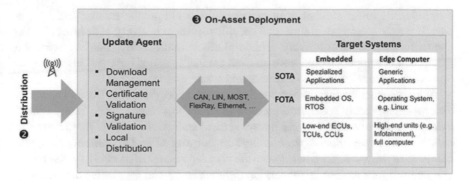

Fig. 5.8 OTA updates – deployment

The local distribution from the update agent to the final target system can occur via a number of different local bus systems, including CAN, LIN, MOST, FlexRay, Ethernet, etc.

The actual target compute node can be a lower-level controller, e.g., an embedded system such as an Engine Control Unit (ECU) or a Transmission Control Unit (TCU) in automotive or any other kind of embedded microcontroller (MCU) or microprocessor (MPU). Alternatively, it can be a high-end edge computer with its own local storage, full operating systems, etc. For lower-level controllers, FOTA (firmware-over-the-air) manages the process of updating the combined embedded OS and embedded applications as a single image (or applying a delta-update strategy to minimize bandwidth), while for higher-level edge computers, SOTA (software-over-the-air) supports more targeted updates of individual applications.

5.7   AIoT AppStores

A particularly interesting application of OTA are AppStores. Pioneered by Apple and Google in the smartphone space, they have proven to offer tremendous value not only in terms of easy-to-use application management, but also with regards to opening up the smartphone platform for external development partners. AppStores for smartphones have unlocked the huge creative potential of millions of developers who are now developing apps for these AppStores, sharing revenues with platform operators.

It seems logical to apply the same principle to other areas, be it AppStores for smart kitchen appliances or electric vehicles. However, there are also some limiting factors. First, with complex and safety-critical systems such as vehicles or heavy equipment, being able to protect them from abuse by potentially malicious external applications is key, and this is not an easy feat. Second, these products usually have a significantly lower number of instances in the field than the smartphone companies. If we are looking at automotive specifically, there is a very high level of fragmentation because almost every vehicle platform is different. This fragmentation means that developers would have an even smaller installed base and high integration costs, making it harder to innovate and create profits.

Achim Nonnenmacher, expert for Software-defined Vehicle at Bosch knows: *In order to significantly accelerate innovation cycles in vehicles from "many years" today to "days or weeks" in the future, vehicle manufacturers and tier 1s need to collaborate on open, non-differentiating software and APIs. This collaboration will help reduce fragmentation, will lower the very high integration effort we see today, and free software engineers to work on differentiating and innovative new features without fearing lock-ins.*

The following is looking at two examples. First, an OEM with closed AppStore for his own vehicle apps. Second, an OEM with an open AppStore for partner vehicle apps.

5.7.1   Example 1: OEM with Closed AppStore

A first step toward an AppStore for vehicles and other physical assets is to only allow apps that are developed directly by the OEM and tier 1 suppliers. Here, we have to differentiate between pure in-car apps (e.g., a new mood control for interior lighting) vs. composite apps with external components, e.g., a smartphone integration.

In the example shown below, a car manufacturer provides a composite app, combining his own AppStore with the smartphone AppStores of Apple, Google, and the like. This type of composite app architecture is required, especially for cases where the app will run partly on the smartphone, partly on the car, and partly in the cloud backend of the OEM.

For example, a user might download a new car app in the smartphone app store. The first time this app is contacting the cloud backend of the OEM, it will determine that a new app component must be installed on the customer's car as well. This could be done automatically and in the background so that the customer is not aware that they are actually dealing with two AppStores.

Once both apps are installed (the one on the smartphone, and the one on the car), the user can then use the apps as one seamlessly integrated app. For example, this could be a new app similar to the Dog Mode app recently introduced by Tesla, which allows a user to control certain features for a dog in the car via his smartphone, e.g., controlling windows and cooling remotely.

This is an interesting scenario for a number of reasons. First, OEMs typically tend to prefer solutions which do not rely on smartphones, since this means losing control over the user. This is why platforms such as Android Automotive are becoming popular, supporting native apps in the car and integrating only with the car's head unit, and not the smartphone. However, by focusing only on on-car apps, the

OEM loses the opportunity for a new customer experience, e.g., by enabling the customer to remotely interact with the car and the dog in it while shopping.

Second, control over the apps in the car app store is critical — from a business — but also from a security perspective. In this scenario, the assumption is that apps in the vehicle app store can only be provided by the OEM or tier 1s. This means that the requirements for the tightness of the sandbox running the applications are not as high, since the OEM has full control over the QA (Quality Assurance) cycle of the app (Fig. 5.9).

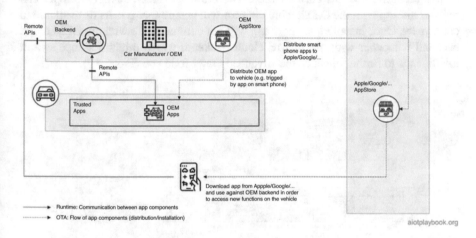

Fig. 5.9   OEM with closed AppStore for vehicle apps

5.7.2   Example 2: OEM with Open AppStore

In this second example, the OEM is actually opening up the car AppStore for external development partners. Let us say the relatively unknown start-up ACME AppDeveloper wants to develop an advanced Dog Mode app, which is utilizing the in-vehicle camera and advanced AI to monitor the dog in the car. Depending on the learning of the AI about the mood and behavior of the dog, different actions can be taken, e.g., modifying the window position, changing the ambient environment in the car, or notifying the owner. Since the OEM does not have an active development partnership with ACME AppDeveloper and similar developers, he has to ensure the following:

- Provide a protected sandbox into which partner applications can be deployed

    - The OEM has to ensure that apps in the sandbox only interact with the car environment via a well-defined set of APIs.

- – Establish a corresponding safety system that ensures that the car is always in a safe state.

- Provide development partners with APIs for sensors such as the in-vehicle camera, as well as selected actuators such as car window control or car ambient control
- Provide OTA-based deployment capabilities for partner apps
- Ensure that development partners can not only deploy specific software but also AI-enabled components

In this scenario, this has been ensured. Therefore, the ACME AppDeveloper can register his app with the OEM, which in turn will ensure that it is made available to car owners. Once installed, the new app will not communicate with the OEM cloud backend but rather with the ACME cloud backend, which might also use swarm intelligence to further enhance the advanced Dog Mode app (Fig. 5.10).

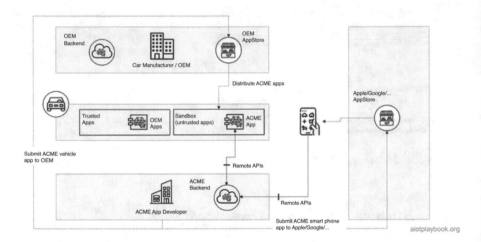

Fig. 5.10 OEM with open AppStore for partner vehicle apps

5.8   Expert Opinion: Nik Willetts, President & CEO of TM Forum

In the following interview with Nik Willetts, President & CEO of TM Forum, different aspects of IoT connectivity and related topics are addressed from the perspective of the telecommunications industry.

Dirk: *Nick, thanks for joining us. Tell us a little bit more about TM Forum. What are you doing? Where are your members, and why is it relevant for AIoT?*

Nik: *Thanks, Dirk. TM Forum is a global consortium of over 800 companies all around the world, largely in the telecoms industry. And that includes the world's leading service providers, software vendors, hyperscalers, system integrators, consultants, and start-up companies. Our purpose is to drive the industry forward through collaboration. Right now, we're focused on transforming the industry's software and operating models to help deliver the agility, time to market and customer experience to unlock growth, at the right cost point. The telcos need to survive in a hypercompetitive market and they foresee huge growth opportunities emerging over the next decade.*

We have been at the forefront for over 30 years of different waves of industry transformation. The current wave is the most fundamental yet, because for connectivity service providers to thrive in the decade ahead they need to digitally transform their business model, operating model and technology stacks. Through our innovation programs, such as our Catalyst projects, we have been exploring the applications of the IoT and combining that with new forms of connectivity. Most recently, we've focused on the application of IoT in combination with AI, edge computation and 5G connectivity in those contexts as well.

So why do we have an interest in AIoT? We believe the next wave of this digital revolution depends on a combination of technologies: elastic connectivity, edge computing, AI, and IoT. This perfect storm of technologies can unleash the true potential of Industry 4.0 and underpin the next wave of digital revolution for society. We know that Machine Learning and Artificial Intelligence will become ubiquitous across those technologies down to the device level. These technologies are going to come from an ecosystem of partners, with expectations over ease of integration, interoperability, and support for new levels of flexibility, agility and new business models. We see that as TM Forum's core competency, in driving collaboration, developing standards, and ensuring that the telecom industry shows up with the right solutions and products, and we want to contribute to making the AIoT vision a reality.

Dirk: *The IoT part in this is not new to telcos. We had M2M, we had telematics, we had IoT, and now we have AIoT. So in the bigger scheme of things, how important is this to the telcos compared to normal communication and the telephone networks, and what are the key market trends?*

Nik: *You're right that mobile technology has been an important element in the first generation of IoT for some time. Today the global market for IoT connectivity is worth approximately $8 billion – not significant when you consider that the services market for telecoms is about $1.6 trillion. However, we see growth both in connectivity revenues for IoT – somewhere between 5% and 7% per annum, and expect the pandemic to accelerate that growth as enterprises bring forward their digital transformation plans.*

I think it is fair to say that the telecom industry has been, at best, a distant partner so far on the IoT journey. We only see a handful of telecom providers with significant IoT divisions, and the immaturity of devices and connectivity technologies have held back deployments. We see significant opportunities for collaboration, between device manufacturers, connectivity providers and end-users. For telecoms providers, the 5G enterprise market is worth at least $700 billion in additional revenue, and much of that will come from use cases which leverage Ml/AI and IoT. It is now critical for connectivity providers to recognize and work *much more closely with end customers, OEMs and software companies. It is not just a case of providing standalone connectivity anymore – we need an ecosystem of technologies to unlock this value.*

Dirk: *If I put myself in the shoes of an OEM building new, smart, connected products, what can I expect from a telco in terms of support for my AIoT deployment, and what do I have to look out for?*

Nik: *With 5G, many leading telco companies are already experimenting with customers. We see pilot projects that span every industry, from fish farming in the Nordics through to health care in the UK. Almost every sizable telecom operator now has deployments and real-world proof of concept projects underway, looking at what's needed from their capabilities and technologies. TM Forum is involved in many of those, including an ongoing project in the manufacturing sector deploying our IoT toolkit and common data model to help manage the friction between Telco, IoT, Cloud and vertical applications. Through these pilots, we see several challenges to navigate.*

The first challenge is what we call Connectivity-as-a-Service — recognizing that for more sophisticated uses of IoT and more advanced devices, you have different connectivity needs at different times and in different locations, and will need to adjust to the available connectivity technologies in that location e.g., 4G, 5G or WiFi. New IoT devices also need to be managed and updated with new software over the air. Those updates, as we already see increasingly more software on cars, have greater bandwidth requirements for short periods of time, along with special requirements over security, latency and privacy. Connectivity has to be flexible and autonomous to support the needs of the IoT device and the required experience.

The second challenge concerns the combination of connectivity with edge computing in regard to rapid AI decision-making. We see AIoT solutions as utilising a combination of intelligence on the device, nearby (at the edge), and centrally in the cloud. Addressing this with a secure, low-latency solution will be key.

The final piece, which we also see as an opportunity, is that security needs and risks are growing. As a regulated industry with substantial cybersecurity experience and control of local networks, telecom operators can provide unique security capabilities, particularly where the processing and use of data can be controlled within a telecom provider's network environment, such as through to an edge computing solution.

Beyond these services, it is also important to note what telcos can offer AIoT use cases. As a global subscription service industry, experience and capabilities such as billing right down to micro-transactions, localization, local market knowledge and skilled workforces, and handling of regulations in local markets are all potentially valuable capabilities. It is important to remember that telcos have global-scale experience delivering complex services and as we have seen in the last 18 months, they are exceptionally resilient to even exceptional levels of demand.

Dirk: *You once said that we need interoperable, autonomous, and open digital eco-*
systems. So what will need to happen in the telecoms industry to actually
achieve this?

Nik: *We believe that a new level of complexity is now coming into play as we com-*
bine and leverage new technologies for more sophisticated and critical use
cases. The first and perhaps the easiest path to manage this is what we call
closed ecosystems. That is where you have a dominant player doing a lot of the
complexity and handling a lot of the integration. We have seen that in our con-
sumer businesses through the evolution of devices from companies such as Apple,
for example, which bring you into a comfortable ecosystem, but ultimately with
significant lock-ins as a consumer.

Lock-in does not work when we get to the complexities of industrial applications
based on IoT – indeed it can directly block innovation, prohibiting you from
embracing newer technologies, experimenting with others, and raising concerns
of customers being held to ransom or, as we have seen in the telecoms industry,
being impacted overnight by costly geopolitical decisions.

So we fundamentally believe that the healthy, sustainable path when it comes to
the next generation of industrial applications, is to build open ecosystems. But
that comes with its own set of challenges compared to closed ecosystems.
Integration, interoperability and transparency are some of the most significant
barriers, and to address those barriers we need open standards. If you don't
have the right standards, if you don't have a common language, definitions, met-
rics, data models APIs and so on, building open solutions becomes very difficult.
That is why our members are creating Open Digital Architecture to help provide
the foundation for open, interoperable and autonomous ecosystems.

So interoperability is key. Fortunately, that is much, much easier today than ever
before. Thanks to advances in software engineering, and the recognition across
industries of the importance of collaboration and standards, it's becoming prac-
tical to deliver the level of interoperability and resilience required for open eco-
systems to thrive.

All of this becomes even more important when we think about the use of AI across
those ecosystems. There's the initial integration of AI across a complex ecosys-
tem of technologies. Then there's the complexity and cost of operating those
technologies. And that's where AI really has a role to play again, not just at the
device level or at a single technology level but actually across devices and tech-
nologies to ensure that the right outcome is achieved for the customer. To do that,
we need to design standards today that are ready for AI use cases of tomorrow.
That will only be possible if all embrace collaboration to deliver the required
standards faster than ever.

Chapter 6
Hardware 101

Dirk Slama

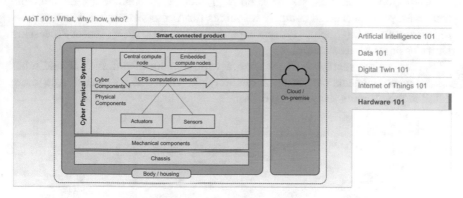

Fig. 6.1 AIoT framework

AIoT hardware includes all the physical components required to build an AIoT product or retrofit solution. For the retrofit solution, this will usually include sensors, as well as edge and cloud (or on-premises) compute resources. Most retrofit solutions will not include actuators. Products, on the other hand, must not only provide IT-related hardware plus sensors, actuators and AI compute resources but also all the mechanical components for the product, as well as the product chassis, body and housing. Finally, both AIoT products and solutions will usually require specialized IT hardware for AI inferencing and AI training. The concepts developed for Cyber Physical Systems (CPS) will also be of relevance here. The following will look at both the AIoT product and retrofit solution perspective before discussing details of the hardware requirements and options for edge/cloud/AI (Fig. 6.1).

D. Slama (✉)
Ferdinand Steinbeis Institute, Berlin, Germany
e-mail: dirk.slama@bosch.com

6.1   Smart, Connected Products

The hardware for a smart, connected product must include all required physical product components. This means that it will include not only the edge/cloud/AI perspective but also the physical product engineering perspective. This will include mechatronics, a discipline combining mechanical systems, electric and electronic systems, control systems and computers.

In the example shown here, all hardware aspects for a vacuum robot are depicted. This includes edge IT components such as the on-board computer (including specialized edge AI accelerators), connectivity modules, HMI (Human-Machine Interaction), antennas, sensors and actuators such as the motors, plus the battery and battery charger. In addition, it also includes the chassis and body of the vacuum robot, plus the packaging.

The cloud or on-premises backend will include standard backend compute resources, plus specialized compute resources for AI model training. These can be, for example, GPUs (Graphics Processing Unit used for AI), TPUs (Tensor Processing Unit), or other AI accelerators.

Setting up the supply chain for such a wide breadth of different IT and other hardware components can be quite challenging. This will be discussed in more detail in the sourcing section (Fig. 6.2).

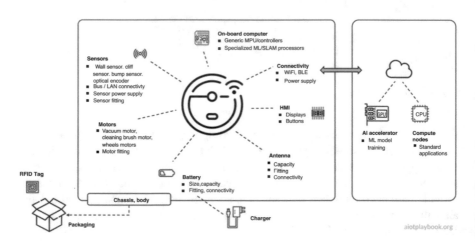

Fig. 6.2 Hardware for smart, connected product

6.2   Smart, Connected (Retrofit) Solutions

For smart, connected retrofit solutions, the required hardware typically includes sensors, edge compute nodes, and backend compute resources (cloud or on-premises). The example shown here is the hardware provided for a predictive maintenance solution for hydraulics components. This complete case study from Bosch Rexroth is included in Part IV.

The hydraulic components include electric and hydraulic motors, tanks, filters, cooling units, valves, and so on. Customers use these components for many different applications, e.g., in manufacturing, process industries, or mining. The hardware components for the retrofit solution include different sensor packs, each specialized for a particular hydraulic component. For example, the sensor package for a hydraulic pump includes sensors to measure pressure and leakage, while the sensor packs for a hydraulic cylinder include sensors for chamber pressure and oil temperature. Since this is a retrofit solution that is sold to many different customers, it is important that each sensor pack has custom connectors that make it easy to attach sensor packs and their different sensors to the corresponding hydraulic component in the field.

Other hardware components provided include an edge data acquisition unit (DAQ), as well as an IoT gateway to enable connectivity to the backend. Backend hardware is not shown in this example, but will obviously also be required.

This is an example of a very mature solution that is designed to serve multiple customers in different markets. This is why the different hardware components are highly standardized. AIoT solutions that are not replicated as often might have a lower level of maturity and a more ad-hoc hardware architecture (Fig. 6.3).

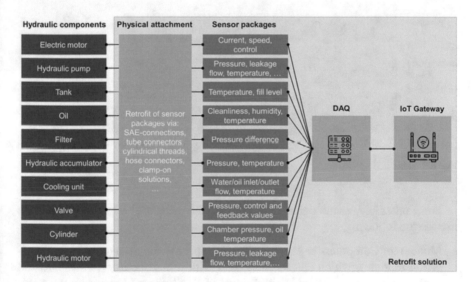

Fig. 6.3 Hardware for smart, connected solution (retrofit)

6.3 Edge Node Platforms

Edge nodes are playing an important role for AIoT. They cover all compute resources required outside the cloud or central data centers. This is a highly heterogeneous space with a wide breadth of solutions, from tiny embedded systems to nearly full-scale edge data centers.

Edge computing is rooted in the embedded systems space. Embedded systems are programmable, small-scale compute nodes that combine hardware and software. Embedded systems are usually designed for specific functions within a larger system.

Today, typical edge node platforms include embedded microcontrollers (MCUs), embedded microprocessors (MPUs), Single Board Computers and even full-scale PCs. While the boundaries are blurry, an MCU is usually a single-chip computer, while an MPU has surrounding chips (memory, interfaces, I/O). MCUs are very low-level, while MPUs run some kind of specialized operating system. Other differences include costs and energy consumption (Fig. 6.4).

	MCUs (Embedded Microcontrollers)	MPUs (Embedded Microprocessors)	PCs/SBC (Single Board Computers)
Cost	€1 - 50	€50 - 500	€100 - 2,500
Use Cases / Applications	Door lock, brake system, fuel injection, air bag	Dialysis machine, parking assistant system, drone control, networking hub	Many different high-end applications, e.g. via containerization
Processor, OS	8 / 16 / 32 bit, simple run-loop / small footprint RTOS	32 / 64 bit, multi-tasking OS / RTOS	x86, 32 / 64 bit, full OS, e.g. Linux
Clock speed / # cores	MHz / single-core	Varies	GHz / multi-core
Power consumption, coolling	Very power efficient (long battery lifetimes), no cooling for compute unit	Moderate energy consumption, active or passive cooling	High energy consumption, active cooling (fans)

aiotplaybook.org

Fig. 6.4 Edge node platforms

Some other key technologies often found in the context of edge node platforms include the following:

- Module (or Computer-on-Module, CoM): a specific function (e.g., a communication module), which can be integrated with a base board via standardized hardware interfaces. Provides high level of flexibility and reuse on the hardware level.
- SoC (System-on-a-Chip): combines multiple modules into a single, tightly integrated chip. Used especially for highly standardized, mass-produced systems, such as smartphones and tablets. For example, a smartphone SoC may contain a CPU, GPU, display, camera, USB, GSM modem, and GPS on a single chip
- ASIC (Application Specific Integrated Circuit): a chip that is custom designed for a specific purpose, e.g., running a mature and hardened algorithm. Provides high performance and low cost if mass-produced but requires a high level of maturity because no changes after production are possible.

- FPGA (Field Programmable Gate Arrays): chips that are programmed using highly efficient but also very low-level, configurable logic blocks. Application logic can be updated after manufacturing.

6.4 Sensor Edge Nodes

Sensor edge nodes are edge nodes that are specifically designed to process sensor data. Most basic sensors actually provide an analog signal. These analog signals are continuous in time, thus consuming a very high bandwidth. They are usually sinusoidal in shape (i.e., they look like a sinus curve). To be able to process and filter these signals, they need to be converted to a digital format. This helps reduce bandwidth, and makes the signals processable with digital technologies. Usually a Digital-to-Analog Converter (DAC) is used to connect an analog device to a digital one. However, before this happens the analog signals are often preprocessed, e.g., using amplification to reduce noise and get a more meaningful signal.

The digitalization of the signal is often done using Discrete Fourier transform (DFT). DFT computation techniques for fast analog/digital signal conversion are known as Fast Fourier Transform (FFT). Based on linear matrix operations, FFTs are supported, for example, by most FPGA platforms.

After conversion to a digital format, the digital signal can now be processed using either traditional algorithms or AI/ML. A discussion on the benefit of edge-based preprocessing of sensor data is provided in the Data 101 section.

Finally, most sensor edge nodes provide some form of data transmission capability to ensure that after preprocessing and filtering the data can be sent to a central backend via an IoT/edge gateway (Fig. 6.5).

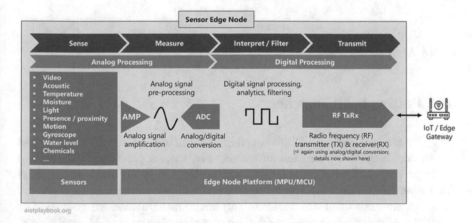

Fig. 6.5 Edge sensor nodes

6.5 AI Edge Nodes

With edge AI becoming increasingly popular, a plethora of specialized edge AI accelerators are emerging. However, with embedded ML frameworks such as TinyML, it is currently possible to execute some basic ML algorithms on very basic hardware. Low-cost IP Cores such as a Cortex-M0 in combination with TinyML can already be used for basic event classification and anomaly detection. Standard MCUs (e.g., an Arduino Nano) can be used to run ML algorithms at the edge for voice recognition and audio classification. Higher-end MCUs even allow for tasks like ML-based image classification. Moving up to full Single Board Computers (SBC) as edge nodes, voice processing and object detection are possible (object detection combines image classification and object localization, drawing bounding boxes around objects and then assigns labels to the individual objects on the image). Finally, SBCs in combination with AI accelerators enable video data analytics (e.g., by analyzing each frame and then drawing a conclusion for the entire video). This is a fast moving space, with many development activities, constantly driving hardware prices down and ML capabilities up (Fig. 6.6).

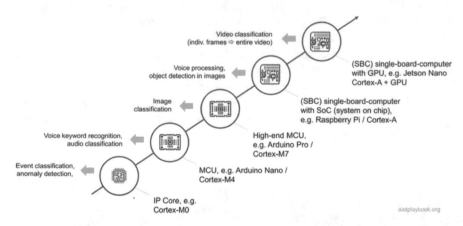

Fig. 6.6 AI edge nodes

6.6 Putting It All Together

Finally, if we are putting all of this together, the following picture is emerging: AI accelerators in the cloud are used for training ever more sophisticated ML models. An important class of AI accelerators are GPUs. Originally used as Graphics Processing Units, GPUs are specialized for the manipulation of images. Since the mathematical basis of neural networks and image manipulation are quite similar (parallel processing of matrices), GPUs are now also often used as AI accelerators. FPGAs are also sometimes considered as AI accelerators since they allow processing very close to the hardware, but still in a way that allows updates after manufacturing. Finally, proprietary solutions are being built, such as TPUs from Google or Tesla's custom AI chip.

Once the model is trained, it can be deployed on the edge nodes via OTA (Over-the-Air-Updates). On the edge node, the model is used on an appropriate hardware platform to perform inference on the inbound sensor data. The loop is ideally closed by providing model monitoring data back to the cloud, in order to constantly improve the model performance (Fig. 6.7).

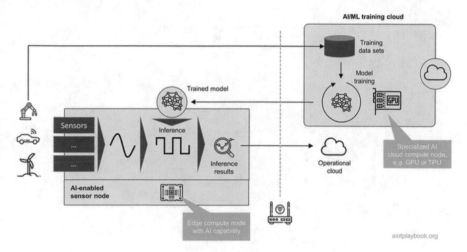

Fig. 6.7 Putting it all together

Part II
Business Strategy

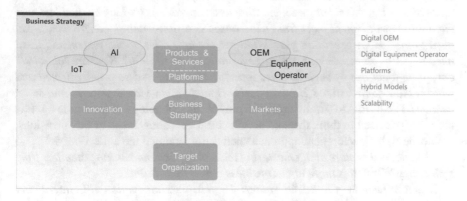

Fig. 1 Business Strategy

Every business strategy must answer the question of how a company should compete in the business areas it has selected, e.g., by applying a cost strategy, a differentiated product or service strategy, or a niche strategy. In the context of AIoT, the business area usually involves physical assets, products, equipment, or appliances. This means that the role of the company will typically be either that of an OEM or an equipment operator, or a combination of the two ('hybrid'). AIoT as a new paradigm will enable both types of companies to create new, digital-enhanced products, solutions, or services. The OEM will become a *Digital OEM*; the equipment operator will become a *Digital Equipment Operator*. For Digital OEMs, the AIoT-enabled digital transformation is often about fundamental changes to the business model. Digital Equipment Operators usually focus more on the digitalization of the operations model. For both, key questions include how to manage innovation and how to define a suitable target organization. Finally, platform-based business strategies have proven to be extremely powerful. Platforms can utilize AIoT to connect to physical products and assets in the field and create a value-added offering. The following expert opinion will shed more light on the AIoT strategy perspective.

Dirk: *Laurenz Kirchner from mm1 consulting. Thanks for joining us. My first question, what is strategy and AIoT to you from the products and services perspective. Do we actually need a business strategy or a dedicated digital strategy?*

Laurenz: *I think both! Or let us say: In this decade — in the 2020s – business strategy cannot exist separated from a digital strategy. When we talk specifically about an AIoT strategy, I believe it is absolutely necessary for any company that either manufactures stuff or operates many assets. These companies will only survive if they transform into organisations that are able to manage connectivity, data and digital services across the stack.*

Dirk: *Okay. So, how do you manage the required innovation at this level?*

Laurenz: *You need to sync your transformation activity on three levels, or three 'plays' as we have shown in the Digital Playbook. I believe you always need to start from a business strategy play, rather than from a technical perspective. So put the business perspective first: what are the priorities for my AIoT transformation? Do I focus on smart products or on automating asset operations? Which critical business capabilities will benefit most from AIoT-based decision-making, etc.? Second, you need to look at the execution play. What are the new roles, tools, responsibilities I need to put in place, such as a data governance organisation? How are 'classical' corporate functions — such as sales, operations, legal, and so on — affected and what is their contribution? How do business models, revenue structures and so on change?* The third AIoT transformation play is the technology execution level. It is here that you need to think through architectural questions, make or buy decisions, define the right development approach such as for example the agile V-model.

Dirk: *So, is this more about mastering technology and innovation potentials from technology? Or is this more about business model innovation?*

Laurenz: *Definitely both! Or rather: The business innovation side interacting with the technology side in a dialectic way. Think 'yin and yang'! A typical situation that we see in client organisations is a lot of push on the technology execution from IT, and this push doesn't really get anchored into the business perspective. Think of a typical use case such as metering: clearly you have to find the right wireless connectivity technology to connect a power meter in the basement of a concrete building. But you also have to think through what is the digital service I want to offer? What is the offering structure? For which part of my service delivery do I actually get revenue? So these things have to work together and you cannot do the one without the other.*

Dirk: *So what about the organizational side? How do I get my organization to support this? What's the target organization? How do I get there?*

Laurenz: *We tend to recommend following a top-down and bottom-up approach. For top-down, you need to enable top management to understand and define the possibilities and the overall objectives that the organization has to pursue, and this takes a lot of education and evangelization at the top level. There's a lot of stuff where executives do not know what they don't know yet.*

Dirk: *Good thing we are writing a Digital Playbook...*

Laurenz: *Exactly. At the same time, you need to define real world showcases and real world use cases; this is the bottom up approach. So you need to find a team — a team of convicts, a team of evangelists — in your organization that are willing to work a little bit, bend the rules and work across functions to make something happen. This can be pure showcase in the beginning, but later on, we recommend truly defining the real business critical use cases to make things happen. This is typically the easiest way you can make an organization follow along and pursue the AIoT transformation.*

Dirk: *Last question, do you see different strategies? Depending on whether you're looking at this from the OEM or product perspective versus the operator perspective?*

Laurenz: *These are definitely different perspectives. What are the challenges of this transformation for a classical OEM? For example, for a maker of power tools, forklifts, or even a maker of cars? Although I do not say the ultimate goal for every one of those organizations is an Equipment as a Service model, the big challenge is still: how do my revenue streams change over time? When I'm used to selling a thousand heavy machinery pieces per year and I know that maybe in ten years I will be selling tens of thousands of microservice digits a week, how do I adapt to this new business model? So this is a typical question for a digital OEM. For a digital equipment operator (let us say a railway operator) clearly the questions are much more: how does this affect my cost base? How will I deliver a certain service or manage my assets in the future? How can I build in things such as connectivity costs and data center costs in the future? How do I arrange for those changes in my own cost base? So very different approaches are needed.*

Dirk: *Thank you.*

Chapter 7
Digital OEM

Dirk Slama

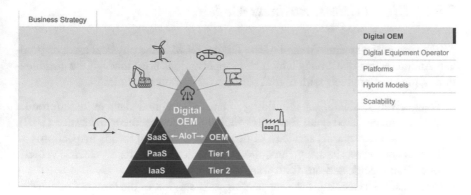

Fig. 7.1 Digital OEM

The Digital OEM combines physical product design, engineering and manufacturing with Software-as-a-Service in order to provide smart, connected products. Artificial Intelligence (AI) and the Internet of Things (IoT) are the two key enablers. This chapter will introduce the concept of the Digital OEM in detail, following again the *why*, *what*, *how* structure from the introduction (Fig. 7.1).

D. Slama (✉)
Ferdinand Steinbeis Institute, Berlin, Germany
e-mail: dirk.slama@bosch.com

© The Author(s) 2023
D. Slama et al. (eds.), *The Digital Playbook*,
https://doi.org/10.1007/978-3-030-88221-1_7

7.1   WHY

The motivation for adopting a digital OEM business model can vary widely. Many incumbent OEMs are seeking ways to build upon their existing business. New market entrants are looking at disruptive new business models enabled by the combination of physical products with AIoT.

While AI and IoT are exciting technical enablers, anybody embarking on the AIoT journey should always start by looking at the "why": Why do this? What is the purpose? And what are the expected business outcomes? From a strategic (and emotional) point of view, the purpose of the AIoT initiative should be clearly articulated: What is the belief? The mission? Why is this truly done?

7.1.1   Digital OEMs: Business Models

At the core of the business model of the Digital OEM is the physical asset or product. An interesting question is which new opportunities arise through the combination of physical products with digital solutions. Examples include:

- Data-driven business, e.g., building on user-generated data or asset/product performance-related data. Examples include usage-based car insurance (UBI), data-driven aftermarket services, or drone-based building facade inspection.
- Digital add-on services, e.g., an optional autopilot service for an electric vehicle, or cooking recipe add-ons for a smart kitchen appliance
- Asset-as-a-Service, e.g., car-seat-heating-as-a-service, or the famous "power-by-the-hour" for Rolls-Royce aircraft engines
- Smart Maintenance, including predictive, preventive and prescriptive maintenance, enabled by deep analytics of asset/machine data via AIoT

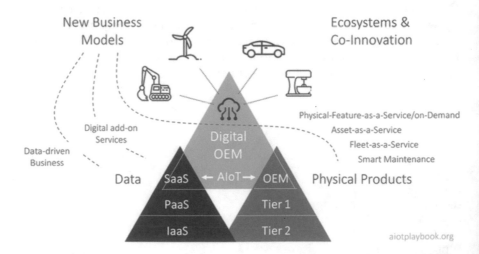

Fig. 7.2 WHY: digital OEM business models

Figure 7.2 shows key elements of two worlds:

- The OEM (Original Equipment Manufacturer) is an organization that makes devices from component parts bought from other organizations. This can be a car maker, a manufacturer of household appliances, or a manufacturer of manufacturing equipment, such as robots or laser cutting tools.
- The suppliers of the OEM are usually referred to as "tier 1", "tier 2", etc., depending on their position in the supply chain
- On the other side, we have the digital ecosystems. Today, large hyperscalers are dominating cloud-based infrastructure (Infrastructure-as-a-Service, or IaaS) and platforms (Platform-as-a-Service). IaaS includes storage, networking, and virtual compute resources. PaaS includes Internet-based tools and middleware for building applications
- Software-as-a-Service (or SaaS) are applications delivered over the internet.
- The digital OEM will combine physical product development with Software-as-a-Service to deliver smart, connected products

7.1.2   Incumbent OEMs: Business Improvements

Especially for incumbent OEMs, the idea of improving existing business by adding digitally enabled solutions is attractive. Generating ARR (Annual Recurring Revenue) via digital services is very interesting, since ARR is seen as a more stable and predictable revenue stream. However, the opportunity to improve existing business – and especially EBIT – with digital solutions as a short-term measure should not be underestimated, since unproven, new business models can have inherent risks and realization of new, ARR-like revenues might take longer than hoped for (Fig. 7.3).

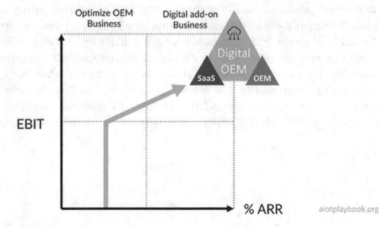

Fig. 7.3 Business outcomes

7.2   WHAT

What can be done with AIoT from the perspective of the digital OEM? Usually, the answer is building smart, connected products. These combine physical products with smartness enabled by AI and connectivity enabled by the IoT. To build smart, connected products, the digital OEM needs to combine product engineering and manufacturing capabilities with edge and cloud software development capabilities (Fig. 7.4).

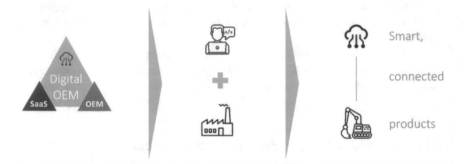

Digital OEMs deliver smart, connected products by combining engineering / manufacturing / logistics capabilities (OEM) with hardware / software / AI development and operations capabilities (often more than in a pure SaaS business)

aiotplaybook.org

Fig. 7.4 WHAT: digital OEM and smart, connected products

7.2.1   Smart, Connected Products: Enabled by AIoT

Smart, connected products usually combine edge and cloud computing capabilities: Edge computing is anything that happens on (or near) the asset/product in the field. Edge computing capabilities are usually dedicated to a single asset/product or sometimes a specific cluster of assets/products operating in close proximity. Cloud computing in an AIoT scenario, on the other hand, can enable insights or functionality that relates to an entire fleet (or "swarm") of assets/products. Consequently, in AIoT, we also differentiate between two types of intelligence: asset/product intelligence vs. swarm intelligence (Fig. 7.5).

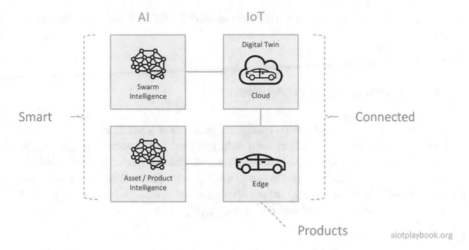

Fig. 7.5 Smart, connected products

7.2.2   Example: Robot Vacuum Cleaner

A good example of a smart, connected product is a robot vacuum cleaner. These products use AI to identify room layouts and obstacles and to compute efficient routes and methods. For example, the robot can decide to make a detour vs. switching into the built-in "climb over obstacle" mode. Another example is the automatic activation of a "carpet boost" mode. IoT connectivity to the cloud enables integration with user interface technology such as smart mobile devices or smart home appliances for voice control ("clean under the dining room table") (Fig. 7.6).

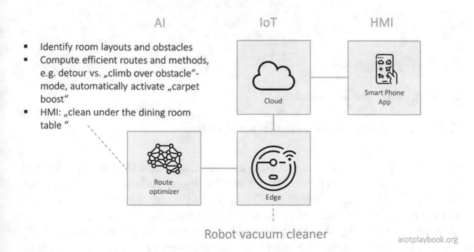

Fig. 7.6 Example: robot vacuum cleaner

The vacuum robot example will be examined in great detail in the product design section.

7.2.3   Example: Kitchen Appliance

Another good example for smart, connected products is a smart kitchen appliance. Here, the intelligence could start with data gathered from users of kitchen appliances in combination with user-generated ratings. These data could be combined to make targeted recommendations (created via AI), e.g., for cooking recipes. A more advanced version of the smart kitchen appliance could also use AI on the product, e.g., for better device control and maintenance (Fig. 7.7).

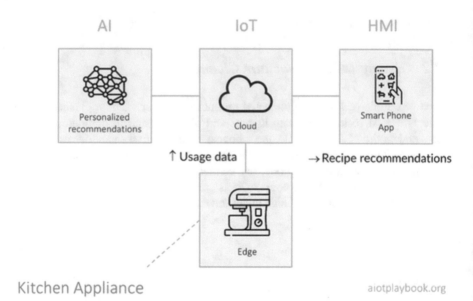

Fig. 7.7   Example: kitchen appliance

7.2.4   Example: Automatic Wiper Control

In this example, AI utilizes images from the autopilot camera to determine the local weather situation. This is then used to automatically convert the wiper speed to the intensity of rain or snow. This is how Tesla is doing it, and it is an area that is also starting to receive the attention of the research community.

What is interesting about this example is that some Tesla customers initially complained that this was not as accurate as other systems using rain sensors. Over time, Tesla was using their Over-the-Air Update (OTA) capabilities to enhance this function, using continuous model improvements and retraining (Fig. 7.8).

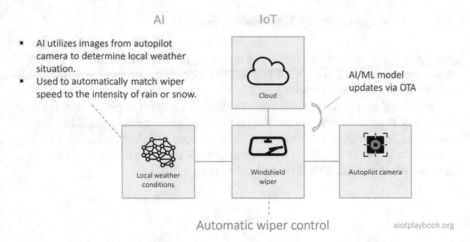

Fig. 7.8 Example: windshield wiper control

7.2.5   Example: Physical Product Design Improvements

Another interesting use of AIoT is for the advanced analytics of product performance, based on data from assets in the field. For example, the team developing the electric motor for the wiper blades from the previous example could use this approach to better understand how their product performs in the field, e.g., at 150 kph on a highway under heavy rain. This information can then be used to improve the next generation of the motor. In this case, it might sometimes not be clear whether we are talking about advanced analytic or real AI (e.g., using ML), but it is still an important use case (Fig. 7.9).

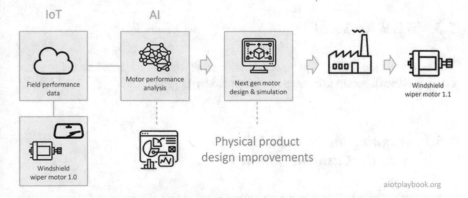

Fig. 7.9 Example: physical product improvements

7.2.6 Example: Smart Tightening Tool

Another example is the smart tightening tool (e.g., the Bosch Rexroth Nexo cordless Wi-Fi nutrunner). This is a type of tool used by industrial customers, e.g., for ensuring the quality of safety relevant joints.

On the tightening tool, AI/ML can be used to control the proper execution of tightening programs (controlling torque and angle for specific combinations of materials). In the cloud, data from fleets of tightening tools can be analyzed to help automatically detect tightening anomalies, classify these anomalies, and make recommendations for handling these anomalies (Fig. 7.10).

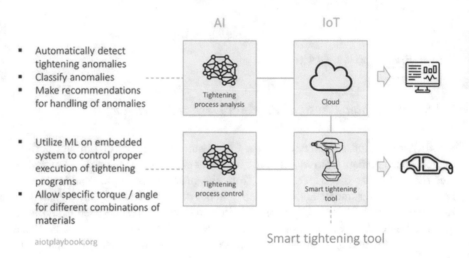

Fig. 7.10 Example: smart, connected tightening tool

7.3 WHY Revisited

Let us revisit the "WHY" perspective with what we have learned thus far about AIoT and the different use cases implemented by Digital OEMs.

7.3.1 Aligning the Product Lifecycle with the Customer Journey

A key feature of AIoT is that it helps align the product lifecycle and the customer journey. In the past, most OEMs lost contact with their products once they left the factory. Although many OEMs try to stay in touch with their customers and support them in the aftermarket, in most cases, the customer relationship was based on service contracts but not a digital relationship. This is of course changing with AIoT, which enables a much higher level of customer intimacy because OEMs can now learn how their products are used in the field and how they are performing. The data obtained and analyzed from the products in the field via AIoT can be augmented with other data, e.g., customer feedback from the Internet.

AIoT also gives the OEM the opportunity to react to what he is learning about his products in the field, by constantly updating existing digital features or even creating new ones, deployed via Over-the-Air Updates (OTA). Naturally, OTA in an AIoT setting will have to support updates of both software and AI models (Fig. 7.11).

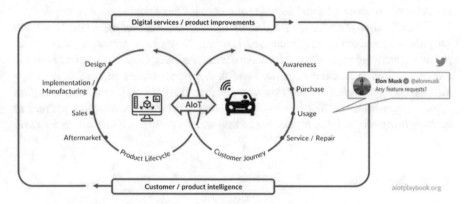

Fig. 7.11 WHY revisited: product LCM and customer journey

This topic was recently discussed by Uli Homann of Microsoft at the BCW.on session with Microsoft CEO Satya Nadella and Bosch CEO Volkmar Denner.

Uli Homann, Corporate Vice President, Microsoft: *The digital feedback loop is essential for successful product development, and OEMs and manufacturers are now also starting to embrace it. For example, we are seeing an increasing number of connected vehicles on the street, which are bringing data into a centralized cloud environment. The cloud is then able to reason over that data and deduce information. Tesla is one of the very famous users of this digital feedback loop already, where they actually use two components. One is the car itself, where it brings information back based upon telemetry, instrumentation, and so forth. So how hard were the brakes being used, if the autopilot is going around the corner? How tight was the corner taken? And then human feedback. Elon is very, very active on Twitter, sometimes very positive, sometimes, to distraction. However, he's very, very active for a very good reason: because he's looking for feedback. One very famous case was people complaining that the Model 3 was taking corners too hard, from their perspective. And so he took the feedback, they compared it with the feedback from the car, and then they made adjustments to the auto drive. And that is truly what we call the digital feedback loop. Because on the one side, you have instrumentation from the car, but you have other channels as well that you bring together and that allow you to start to really think about the lifecycle of the customer journey, the customer buying the car, finding the right car, servicing the car and those kinds of things, and bringing all that data, all of this awareness back into the engineering cycle from design to manufacturing, to the sales and after sales, after market opportunities, etc. Bringing this together in an intelligent way based upon data, utilizing AIoT, is truly the key piece here. The last dimension of making this happen are open platforms, both from an approach to software development as an open ecosystem, with open tools and a lot of open source in the cloud, and also open standards coming together not only in the cloud but also extending this reach into the car. The resulting programming model has platform capabilities underneath that are derived from the Cloud, and optimized for the car. Making this happen consistently will not only allow us to enable AIoT in the cloud but also bringing cloud into the car or into the manufacturing capability. I think the digital feedback loop, the platform tooling and then bringing it into a consistent end-to-end perspective truly will help ensure that we can get digital services at your fingertips. Again, Microsoft is part of an open ecosystem here. We are working together with Bosch and other players to actively bring this to bear, to real life so that we can truly drive this vision forward.*

7.3.2 Benefits

The benefits of this approach are manifold, including shorter time-to-market, improved differentiation, improved sales (including recurring revenues), improved customer experience, and consequently improved customer loyalty (Fig. 7.12).

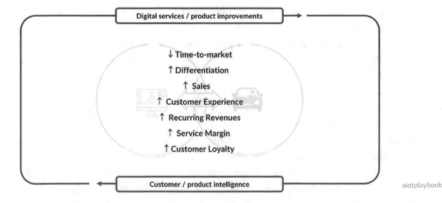

Fig. 7.12 WHY: benefits

7.4   HOW

Now let us take a closer look at how the Digital OEM must go about implementing this with AIoT. This will include a discussion of key design decisions, technical constraints, and considerations for execution and delivery.

7.4.1   Key Design Decisions

From the product manager's perspective, a key question in the future will be – for each feature – whether this feature should be implemented in hardware, software, or AI/data, or combinations thereof. Implementing a feature in hardware (including HMI, processing, etc.) will have an impact on usability (for example, sometimes it will still be preferable to activate a feature via a physical control) but also on engineering and design complexity. Implementing the same feature completely in software (e.g., as a feature activated via a smart app) can often mean a lower cost of delivery (no manufacturing/supply costs beyond the initial development) and also means that the feature can be updated via OTA in the future. Finally, if the feature can be implemented virtually, then the next big question is whether it should be implemented as a set of hard-coded rules (software development) or as a data-centric AI function that uses inference to make a decision based on its training.

The decision to use AI, Software, or Hardware for a specific feature will have two main implications: first, the quality of the User Experience (UX), and second, the required technology pipeline to deliver the feature (Fig. 7.13).

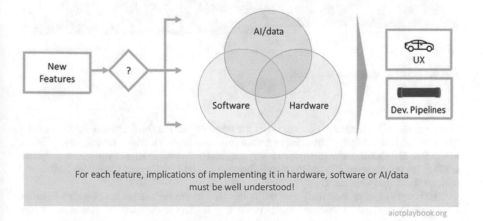

For each feature, implications of implementing it in hardware, software or AI/data must be well understood!

aiotplaybook.org

Fig. 7.13 HOW: key design decisions

7.4.2   Considerations for Execution and Delivery

For the digital OEM, execution and delivery will require a holistic view, including business model, leadership & organization, sourcing and co-creation, User Experience (UX) and Human/Machine Interfaces (HMI), data strategy, AIoT architecture, DevOps, Digital Trust and Security, Quality Management, Compliance and Legal, Productization and Sales; and how AIoT will impact them.

What is usually less relevant for the Digital OEM are aspects such as retrofit (assuming the approach here will be a line-fit approach), site preparation and rollout. These are all important aspects for the Digital Equipment Operations, which will be discussed next (Fig. 7.14).

Digital OEM	Physical Product	Digital Services	
Relevance Area		AI	IoT
●●● Business Model	✓	✓	✓
●●● Leadership & Organization	✓	✓	✓
●●● Co-Creation & Sourcing	✓	✓	✓
●●● Engineering & Manufacturing	✓		
●●● UX & HMI	✓	✓	✓
●●● Data Strategy		✓	✓
●●● AIoT Architecture		✓	✓
●●● AIoT DevOps		✓	✓
●●● Digital Trust, Security		✓	✓
●●● Quality Management	✓	✓	✓
●●● Compliance and Legal	✓	✓	✓
●●● Productization	✓	✓	✓
●●● Sales	✓	✓	✓
●○○ Fit into existing processes	✓	✓	✓
○○○ Retrofit / site preparation / rollout			

aiotplaybook.org

Fig. 7.14 HOW: execution and delivery

Chapter 8
Digital Equipment Operator

Dirk Slama

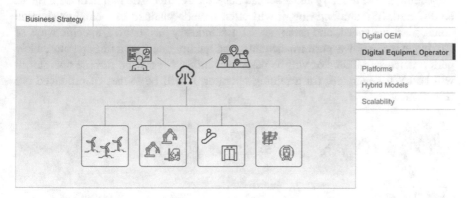

Fig. 8.1 Digital equipment operator

The Digital Equipment Operator utilizes AIoT to optimize how they operate physical assets or equipment. Goals often include asset performance optimization and process improvements. Examples of Digital Equipment Operators include manufacturers, electricity grid operators, railroad operators, and mining companies. This chapter introduces the concept of the The Digital Equipment Operator in detail, again following the *why*, *what*, *how* structure from the Introduction (Fig. 8.1).

D. Slama (✉)
Ferdinand Steinbeis Institute, Berlin, Germany
e-mail: dirk.slama@bosch.com

© The Author(s) 2023
D. Slama et al. (eds.), *The Digital Playbook*,
https://doi.org/10.1007/978-3-030-88221-1_8

8.1 WHY

The motivation to become a Digital Equipment Operator can be manifold. A good starting point is to look at OEE (Overall Equipment Effectiveness), or whatever the equivalent in the specific industry. OEE measures the performance of an asset compared to its full potential. OEE quantifies the utilization of manufacturing resources (including physical assets, time, and materials) and provides an indication of any gaps between actual and ideal performance. OEE is often calculated based on the following three metrics:

- Availability, e.g., asset up-time
- Performance, e.g., system speed
- Quality, e.g., levels of defects

Depending on the industry and asset category, the detailed calculation of OEE might be different. In manufacturing, it will often include planned vs. actual production hours, machine speed, and scrap rates. Each industry has its own, specific ways of looking at availability, performance rate, and quality rate. For a rail operator, a key performance rate indicator will be passenger miles; for a wind turbine operator, it will be kWh generated; for a mining operator, it will be tonnes of produced ore (Fig. 8.2).

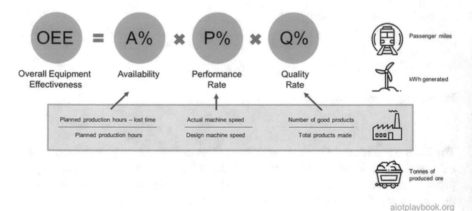

Fig. 8.2 Why

8.2 WHAT

AIoT can support the Digital Equipment Operator in many ways. Smart, connected solutions can help improve OEE through better visibility, advanced analytics, and forecasting. Asset Performance Management (APM) aims at taking a holistic view of asset performance, utilizing data, and insights generated via AIoT. Availability can be improved with AIoT-based predictive, preventive, and prescriptive maintenance. The performance rate can be improved with AIoT-generated insights for process optimization. Quality management can be supported by AIoT-enabled quality control mechanisms, e.g., optical inspection. Advanced analytics and AI can also help improve the quality rate (Fig. 8.3).

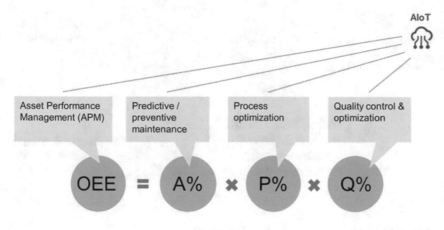

Fig. 8.3 What

8.2.1   Example: Escalator Operator (Railway Company)

The first example we want to look at is a railway company that is operating escalators at its train stations. For a large railway company, this can mean a fleet of thousands of escalators in a wide geographic range. Most likely, the escalator fleet will be highly heterogeneous, including products from many different vendors. Reducing downtimes will be important for customer satisfaction. In addition, obtaining improved insights into the escalators' operational health status can also help reduce operations and maintenance costs.

For the rail company, escalator monitoring has to work for the entire fleet and must provide seamless integration with the facility management operations system. Getting all of the different suppliers on board to agree on a common solution will be impossible. The logical consequence is to design a solution that can be applied to existing escalators in a retrofit approach, potentially even without support from the escalator vendor/OEM.

A good technical solution in this case is to utilize AI-based sound pattern analysis: sound sensors attached to the escalator provide data which can be analyzed either using an on-site edge node, or centralized in the cloud. AI-based sound pattern analysis provides insights into the current state and can even predict the future state of escalator performance.

Once a problem is identified or forecasted, the train station operations personnel or the facility management organization is provided with this information and can take appropriate action (Fig. 8.4).

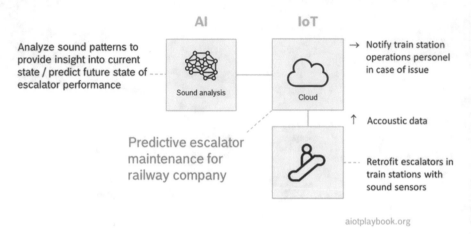

Fig. 8.4 Example: escalator operator (railway company)

8.2.2 Example: School Bus Fleet Operator

Another good example of a Digital Equipment Operator is a school bus fleet operator, utilizing AIoT to provide a platform that offers shuttle services for schools. Instead of using a fixed bus network and fixed bus schedule, the service utilizes AIoT to offer a much more on-demand service to students. Instead of using fixed bus stops, virtual bus stops are introduced that can change during the day, depending on demand. Students can use a smartphone app to request a ride to and from the school. Shuttle buses are equipped with an on-board unit to provide bus tracking and AI-based in-vehicle monitoring. The platform in the backend utilizes AI to optimize the pick-up order and routing of the shuttle buses (Fig. 8.5).

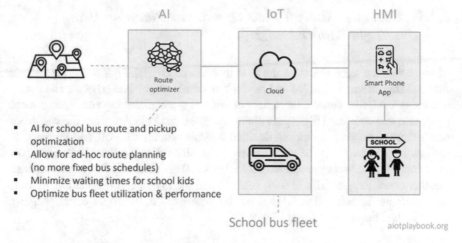

Fig. 8.5 Example: school bus fleet operator

This example will be discussed in more detail in the Sourcing Chapter. The figure following shows an example of how the routes for multiple vehicles can be optimized to support multiple stops on a dynamic route (Fig. 8.6).

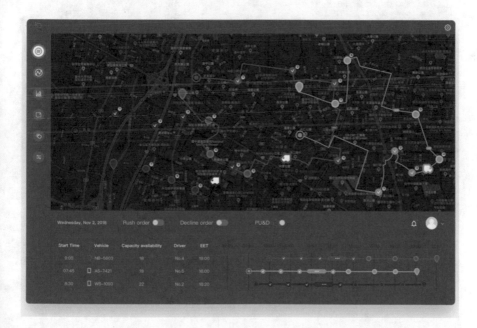

Fig. 8.6 UX for school bus shuttle

8.2.3 Example: Aircraft Fleet Operations Planning Using a Flight Path Optimizer

Modern airlines were amongst the first to become Digital Equipment Operators, first utilizing telematics, M2M and now IoT in combination with advanced analytics and today's AI. Managing a large fleet of aircraft is a challenging task. One critical process in this context is flight path planning. The flight path describes the way from one airport to another, including detailed instructions for take-off and landing, as well as the way between the two airports. From the airline's point of view, the two most important aspects are safety and fuel costs. The latter requires inputs such as weather conditions, overflight fees, fuel costs at the origin and destination, as well as aircraft performance data. Based on this information, the flight path optimizer can calculate the optimal route (Fig. 8.7).

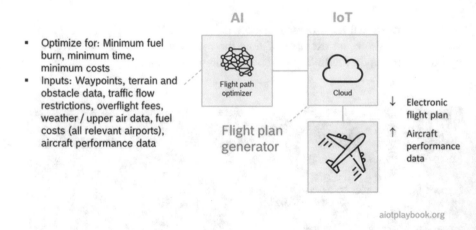

Fig. 8.7 Example: airplane fleet operations planning using flight path optimizer

8.3 HOW

There is no one-size-fits-all answer to becoming a Digital Equipment Operator. This section looks at a generic Solution Lifecycle, as well as considerations for execution and delivery (analogous to the previous section).

8.3.1 Solution Lifecycle

For some Digital Equipment Operators, there will be a central AIoT solution that is at the core of their fleet operations. The airline's fleet planning system might be such a core application. However, very often, Digital Equipment Operators will find that the solutions they require are on the long tail of the AIoT chart (see discussion on the long tail in AIoT 101). This means that they are looking at building multiple, specialized solutions, that need to be constantly enhanced and adapted. This can be supported by a *measure/analyze/act* approach. The AIoT in High-Volume Manufacturing Network case study provides a good example for this approach (Fig. 8.8).

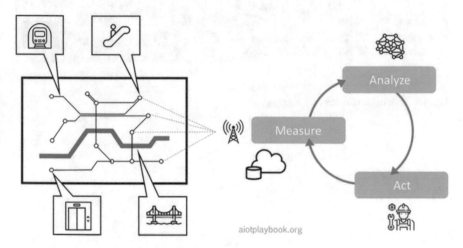

Fig. 8.8 How

8.3.2 Considerations for Execution and Delivery

For the Digital Equipment Operator, execution and delivery will require a different perspective than for the Digital OEM. While any investment will have to be justified by a matching business case, the overall business models tend to be much more straightforward. Similarly, leadership and organization are important but probably not as challenging. Other aspects, such as sourcing, UX, DevOps, compliance and legality, and productization, will be less important compared to the Digital OEM. Data strategy, on the other hand, will be key, especially if a multitude of potentially heterogeneous data sources — sensors and enterprise systems — will have to be integrated. Finally, the Digital Equipment Operator will have to focus on fitting the new AIoT solutions into existing business processes. And in order to get there, asset retrofit, site preparation, and rollout management will be key prerequisites (Fig. 8.9).

Digital Equipment Operator		Physical Product	Digital Services	
Relevance	Area		AI	IoT
● ○ ○	Business Model		✓	✓
● ● ●	Leadership & Organization		✓	✓
● ● ●	Co-Creation & Sourcing		✓	✓
○ ○ ○	Engineering & Manufacturing			
● ○ ○	UX & HMI		✓	✓
● ● ●	Data Strategy		✓	✓
● ○ ○	AIoT Architecture		✓	✓
● ● ○	AIoT DevOps		✓	✓
● ● ●	Digital Trust, Security		✓	✓
● ● ●	Quality Management		✓	✓
● ○ ○	Compliance and Legal		✓	
○ ○ ○	Productization			
○ ○ ○	Sales			
● ● ●	Fit into existing processes		✓	✓
● ● ●	Retrofit / site preparation / rollout	✓	✓	✓

aiotplaybook.org

Fig. 8.9 Execution and delivery aspects

Chapter 9
Platforms

Dirk Slama

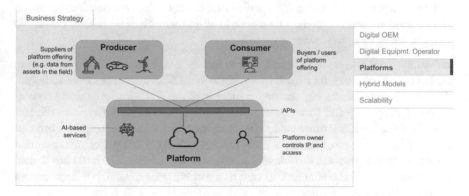

Fig. 9.1 AIoT-enabled platforms

From Airbnb to Amazon and Uber, digital platforms have disrupted many industries, including travel, retail, music and others. From an AIoT point of view, we are interested in digital platforms that connect to physical assets and products. An AIoT platform operator does not necessarily have to manufacture or operate the physical assets. This is why platforms are covered independently of the Digital OEM and Digital Equipment Operator roles introduced earlier, even though the approaches can of course also be combined. This chapter provides an overview of the different concepts and looks again at the *why*, *what*, and *how* of platforms (Fig. 9.1).

D. Slama (✉)
Ferdinand Steinbeis Institute, Berlin, Germany
e-mail: dirk.slama@bosch.com

© The Author(s) 2023
D. Slama et al. (eds.), *The Digital Playbook*,
https://doi.org/10.1007/978-3-030-88221-1_9

9.1 WHY

Successful digital platform businesses create network effects that scale globally, creating huge revenues and profits without having to invest in a physical infrastructure. Since digital platforms rely on an ecosystem of external producers and consumers, the only limit to scale is the size and value of the ecosystem. Uber relies on independent drivers, as well as people in need of rides. Airbnb does not own hotels or apartments, but has grown a multibillion business as a neutral broker, taking a cut off every transaction. Generally, it seems that there are two distinct motives for becoming involved in a digital platform business:

- Winner takes all: Many platform businesses are fairly dominant in their own domain, making them either extremely profitable or an extremely interesting growth investment
- Underdogs team up: Often, companies that cannot achieve a dominant position in a platform market on their own seek to align themselves with other players to try to catch up to the leader

Many industrial players have been trying to imitate the success of B2C platforms in the last couple of years, motivated by the scale these B2C businesses have achieved. However, it seems that a key reason for success of the B2C platforms was a combination of simplicity and focus. Many B2B platform scenarios suffer from the fact that industrial solutions often tend to be much more complex, and require a broader approach. For example, creating a platform that brokers holiday apartments cannot be compared to a platform that brokers automotive sensor data. The latter has to deal with more complex stakeholders, as well as a much higher diversity of data sources (there can be dozens of different sensors on a car, and they will differ from model to model). Since many AIoT use cases are of an industrial nature, there also seems to be an opportunity here to address this.

9.2 WHAT

The basic concept of a multi-sided business platform has been well described by G. Parker et al. in *Platform Revolution: How Networked Markets Are Transforming the Economy* [10]. The platform provides the infrastructure (e.g., an appstore) and brings together producers and consumers. The producers are creating the platform content (e.g., apps in the appstore), while the consumers are buying or using it (e.g., by downloading apps to their smartphones).

Three paradigm shifts are described as key for moving toward a platform business model. First, the move from *resource control to resource orchestration*. Traditional companies have tangible assets on their own balance sheets, e.g., real estate, factories or mines. Platform businesses are based on a less tangible asset, the ecosystem of providers and consumers. The value of Airbnb is the large community of holiday home owners and seekers. Second, *from internal optimization to external interaction. Again, traditional companies are focusing on internal activities to create and sell products or services. Platform businesses focus on value creation by building external ecosystems. Third, "from a focus on customer value to a focus on ecosystem value". Traditional companies focus on the lifetime value of their customers. Platform companies focus on creating network effects between their customers.*

AIoT-enabled platforms include physical products or assets as a key part of the ecosystem that creates the platform network effect. For example, this can mean that physical products provide data, which is then consumed by platform customers - either human users, or other physical products. The IoT enables connectivity, either in a producer or a consumer role or both. AI can support the producer in creating a meaningful offering. Equally, it can support the consumer in making use of the platform, e.g., by processing data from a platform in a customer-specific way. Alternatively, AI can be applied by the platform operator to create swarm intelligence that benefits from multiple data producers.

9.3   HOW

The authors of the *platform revolution* provide three recommendations for building a platform: magnetism (producers and consumers must attract each other), user-generated content, and implicit creation of value by the platform owner. Applying this to an AIoT-enabled platform, this means:

- Magnetism: The AIoT-enabled platform must find a match between consumers and providers that creates this magnetism. This will heavily depend on the type of physical assets or products involved, and the supported use case.
- User-generated content: For example, sensor data from assets in the field
- Implicit creation of value: For example, a swarm intelligence that combines the data from multiple sensors, e.g., to create a real-time map or road conditions

Since this is difficult to generalize without being too generic, let us take a look at a concrete example in the following.

9.4   Example: Parking Spot Detection (Multi-Sided Business Platform)

This is an example of a multi-sided business platform enabled by AIoT: cars equipped with ultrasound sensors can detect available parking spots as they are passing by them. These data are collected in a centralized platform and monetized, e.g., via a find-a-free-parking-spot app. Multiple OEMs might provide parking data to the platform operator, which integrates, consolidates and markets the data (Fig. 9.2).

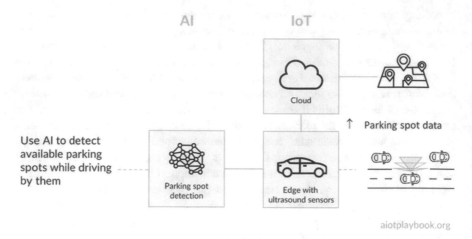

Fig. 9.2 Example: parking spot detection (multi-sided business platform)

9.5 Challenges

As mentioned earlier, building successful platform business models in industrial environments - or environments involving highly complex products such as cars - is not easy. Often, this is because in these environments we are facing a mixture of technical complexity, stakeholder complexity, and legacy (physical assets, products and equipment in the field are often suffering particularly badly from high levels of heterogeneity, because of their often long lifetimes).

A number of data marketplaces have emerged in the last couple of years that focus on bringing together OEMs and after-market customers, e.g., in automotive. The challenge here is manifold. First, OEMs do not always have an interest in making their data available, not even for payment. Second, the question is how to integrate - through basic APIs accessible over the internet, or through custom hardware deployed on the vehicles (which allows better integration, but increases costs). There are also some startups that are providing completely generic marketplaces for sensor data. The problem here is that they are often too generic, making it difficult for users to truly find a relevant offering (missing *"magnetism"*).

Another potentially interesting area for AIoT-enabled platforms is industrial AppStores, or AppStores for complex consumer products such as cars or kitchen equipment. The example from before would not be possible without such an AppStore. However, there are at least two challenges here. First, the number of relevant consumers of the apps is most likely much smaller than in a smartphone AppStore, thus making it more difficult to build profitable apps. Second, the OEM would have to provide the app developer access to APIs, to get access to sensors such as the ultrasound sensors in the previous example. While many smartphone vendors are making increasingly more sensor APIs available to app developers, this might still take a while in industries such as automotive, simply because any security problems with the OEM's app sandboxes could have potentially cata-strophic consequences. An interesting step along this journey could be app stores which are only accessible to trusted partners, as we have described in the co-creation section.

Therefore, while AIoT-enabled platform businesses are certainly not straightfor-ward, it will be interesting to observe how the industry will approach this, and who will be the first players to succeed in their areas. It seems fair to say that smartphone app store players have already shown how to do this (using their own form of AIoT), and others will eventually follow in their domains.

Chapter 10
Hybrid Models

Dirk Slama

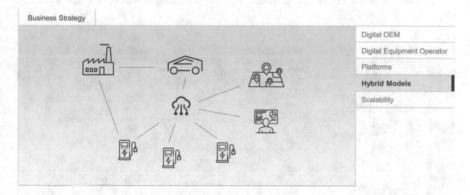

Fig. 10.1 Hybrid models

Hybrid models that combine aspects of the Digital OEM and the Digital Equipment Operator are more often the norm than the exception. Nevertheless, differentiating between these two roles can be very helpful for understanding many of the different concepts associated with them. This chapter looks at hybrid models in more detail, following again the *why*, *what*, *how* structure from the Introduction (Fig. 10.1).

D. Slama (✉)
Ferdinand Steinbeis Institute, Berlin, Germany
e-mail: dirk.slama@bosch.com

© The Author(s) 2023
D. Slama et al. (eds.), *The Digital Playbook*,
https://doi.org/10.1007/978-3-030-88221-1_10

10.1 WHY

In many cases, companies will seek to create an integrated business model that combines the OEM and operator roles. For example, an electric vehicle manufacturer might also choose to own and operate its own network of fast charging stations (such as Tesla with its network of supercharger stations). In this case, combined KPIs are likely to include revenue and usability.

Alternatively, the OEM may choose an Asset-as-a-Service business model, which also means that it will play a hybrid role. In this case, the hybrid model will be based on a combination of typical OEM and operator KPIs, e.g., including both revenue and OEE.

Another example of a hybrid model is the Productized Retrofit Solution, e.g., a productized predictive maintenance solution. Here, the KPIs will combine revenue with costs for customer-specific modifications to the solution (Fig. 10.2).

Hybrid Model	Examples	OEM/Operator KPIs
Integrated Business Model (OEM & Operator)	• EV manufacturer with own network of fast charging stations • OEM with Asset-as-a-Service business model	• ↑ Revenue, ↑ usability • ↑ Revenue, ↑ OEE
Productized Retrofit Solution	• Predictive maintenance solution with own sensor packages • Track & Trace solution with productized tracking sensors	• ↑ Revenue, ↓ costs for customizations • ↑ Revenue
Other kinds of digital/physical offerings	• Drone-based natural disaster area inspection for insurance companies	• ↑ Revenue

Fig. 10.2 Why

10.2 WHAT

The definition of *What* exactly constitutes a hybrid model will heavily depend on the specifics of the product or solution. The table following compares the typical aspects of the Digital OEM and the Digital Equipment Operator, including aspects such as typical customers, offering, positioning of the assets on the balance sheet, typical KPIs, level of standardization, etc. A hybrid model might differ in any of these dimensions, or provide a combination thereof (Fig. 10.3).

	Digital OEM	Digital Equipment Operator	Platforms	Hybrid Models
Customer:	B2C or B2B	Usually B2B	Providers / Consumers	
Offering:	Smart, connected products	Smart, connected solutions	Data or services	
Organization:	Product organization	Project/program organization	Digital product org.	
Balance sheet:	Sold physical products **not** included in own balance sheet	Physical assets/equipment recorded under PP&E (property, plant, and equipment)	Not assets	
KPIs:	Useability / ease-of-use, EBIT, revenue (one-time sales, subscriptions)	Efficiency, cost reduction, quality improvement	Platform-usage related	
Physical Product / Equipment:	Specifically designed and manufactured to support AIoT product offering (**line-fit**)	**Retro-fit** to existing, often heterogeneous assets	Integration via **APIs**	
Standardization:	Productization (packaging, installation, documentation, training, support)	Industrialization (project documentation / operating procedures, operational support)	Highly standardized APIs	
Delimitation:	Multiple, fully isolated users (multi-tenancy)	Single installation with holistic fleet perspective (single tenant)	Multi-tenancy, data-space concepts	

aiotplaybook.org

Fig. 10.3 Comparison

10.2.1   Example: Predictive-Maintenance-as-a-Service

A good example of a hybrid model is the Predictive-maintenance-as-a-Service solution described in detail in the case study from Bosch Rexroth. One product category offered by the company are hydraulic components, including hydraulic motors, pumps, tanks and filters. These hydraulic components are used in many different applications, including manufacturing, mining and off-road vehicles. In order to offer customers of these hydraulics products an improved maintenance solution, Predictive-maintenance-as-a-Service for hydraulic components was developed. The main issue here is that it turned out that the algorithms for detecting anomalies related to the hydraulic components differ from application to application. To address this, Rexroth had to establish a setup that allowed them to standardize the offering as far as possible, and provide customer-specific customizations as efficiently as possible. This is a good example of a productized retrofit solution as per the table preceding (Fig. 10.4).

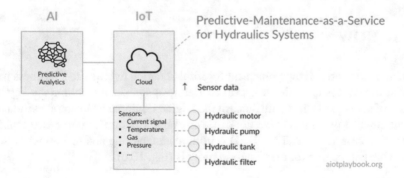

Fig. 10.4 Example: predictive maintenance

10.2.2 Example: Drone-based Building Facade Inspection

Another good example of a hybrid model is the Drone-based Building Facade Inspection developed by TUEV SÜD. Again, this is described in more detail in the case study. This solution utilizes drones equipped with cameras, thermal scanners, and LIDAR (laser scanners) to create a detailed scan of the facades of buildings. On the drone, AI is used for flight path control, collision avoidance, and position calculation. In the TÜV Cloud, AI is used to detect anomalies such as concrete cracks, concrete spalling, corrosion, etc. The customer gets a detailed report about any potential problems with the facade that need to be addressed as part of the building maintenance process. In this case, TÜV SÜD is both the OEM and operator of the solution (Fig. 10.5).

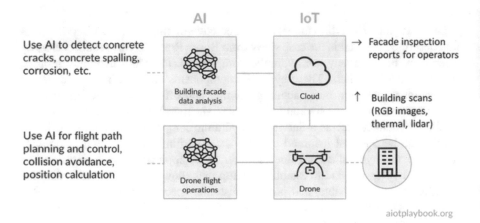

Fig. 10.5 Example: drone-based building façade inspection

10.3 HOW

Again, there is no common blueprint for implementing hybrid AIoT business models. However, it is clear that a hybrid model must somehow combine the key processes of the Digital OEM with those of the Digital Equipment Operator, as indicated by the figure following. As we noticed, for example, in the Drone-based building inspection case study, TUEV SUED is both manufacturing and operating the solution. This will be true for many hybrid models (Fig. 10.6).

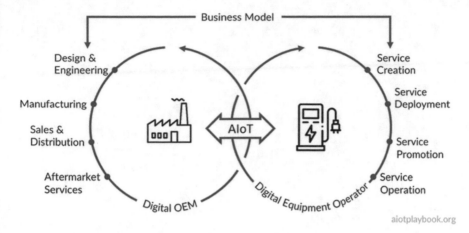

Fig. 10.6 Hybrid models: HOW

Chapter 11
Scalability

Dirk Slama

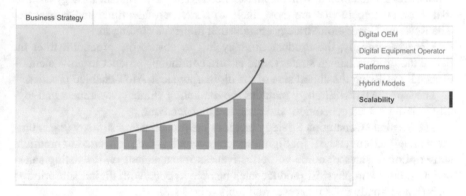

Fig. 11.1 Scalability

The ultimate goal of the business strategy is to ensure that the business can be scaled up to the level that matches the business objectives. This is usually a step-by-step process, involving exploration, acquiring early adopters, and then continuously growing the business. Which of the methods that have worked for successfully scaling purely digital businesses can be adopted by AIoT-enabled businesses? What are the pitfalls of scaling up a digital/physical business? (Fig. 11.1).

D. Slama (✉)
Ferdinand Steinbeis Institute, Berlin, Germany
e-mail: dirk.slama@bosch.com

© The Author(s) 2023
D. Slama et al. (eds.), *The Digital Playbook*,
https://doi.org/10.1007/978-3-030-88221-1_11

11.1 Understand Strategy Requirements

The first important step is to understand the key elements required for the commercialization and scalability strategy. These elements will be different for the Digital OEM vs the Digital Equipment Operator, as will be discussed in the following.

11.1.1 Digital OEM: Strategy for Smart, Connected Products

In order to successfully commercialize smart, connected products at scale, the Digital OEM will need to address three strategic elements: product strategy, go-to-market strategy, and revenue generation strategy. Product strategy ensures that the product has an excellent fit with the market needs. The go-to-market strategy ensures that those customers who are most likely to benefit are identified and persuaded. The Revenue Generation Strategy ensures that money is coming in.

More specifically, the Product Strategy has to ensure the product/market fit, define the product launch strategy, and ensure continuous product improvements – especially utilizing the digital side of the digital/physical, AIoT-enabled product.

The Go-to-Market Strategy includes the marketing strategy, awareness and loyalty programs, lead generation, and retention management.

The Revenue Generation Strategy includes the monetization strategy (e.g., starting with a freemium model for digital services, which is then converted to premium subscriptions), sales resource and effectiveness management (will existing salespeople focused on physical product sales be able to cope with digital subscription sales?), and finally sales processes and tools (Fig. 11.2).

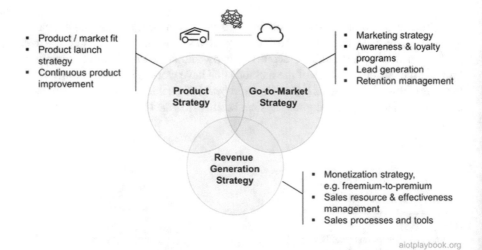

Fig. 11.2 Commercialization and scalability strategy for smart, connected products

11.1.2 Digital Equipment Operator: Strategy for Smart, Connected Solutions

The Digital Equipment Operator will often focus on creating smart, connected solutions to optimize internal processes. Consequently, commercialization is not so much of relevance here. Instead, continuous optimization is key – usually related to the different elements of OEE (overall equipment effectiveness): availability, performance rate, and quality rate. Consequently, the key elements of the best matching strategy include Solution Strategy, Rollout Strategy, and OEE Optimization Strategy.

The Solution Strategy includes a strategy to match the solution to internal demand, a solution launch strategy, and a strategy to continuously improve the solution itself.

The Rollout Strategy usually includes a strategy for site preparation, a retrofit program (how to retrofit 2000 escalators at 200 train stations?), and an internal awareness and adoption program (how to convince the internal stakeholders to actually use the solution) (Fig. 11.3).

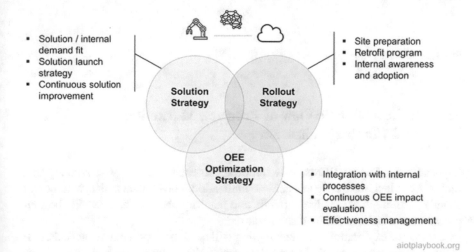

Fig. 11.3 Commercialization and scalability strategy for smart, connected solutions

11.2 Clearly Define Your Focus Areas

The first step toward ensuring scalability of an AIoT-enabled product business or internal optimization effort is to clearly define the focus areas: is this about optimizing core business processes by integrating them with intelligence from assets in the field? If so, which ones: marketing, sales, operations, manufacturing? Is the focus on revenue and profitability, or on OEE (Overall Equipment Effectiveness)? Is this about new or improved user experience, e.g., by adding a digital experience to a previously purely physical product? Or is this about disrupting channels, e.g., by opening up a new channel that could even be competitive with one of the existing channels. Or is it even about creating a completely new business, e.g., by creating a new digital/physical product category? Understanding and clearly articulating the focus area should be the first step of every AIoT-enabled digital transformation effort (Fig. 11.4).

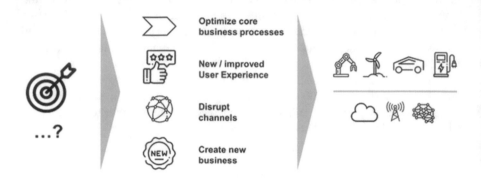

Fig. 11.4 Clearly define focus areas

11.3 Take a Holistic View of Product, Marketing and Commercialization

Especially for Digital OEMs, it is important to establish a holistic strategy that includes product, marketing and commercialization. The product and its market will usually go through exploration, growth, and maturity phases. These need to be supported by marketing and commercialization.

During the exploration phase, marketing will need to support market need assessment and market validation. During the growth phase, awareness and visibility, as well as lead generation, needs to be supported. Finally, when moving to the maturity phase, support for customer loyalty and retention will become more important.

From the point of view of the commercialization strategy, in the exploration phase the analysis of economic feasibility plays a key role, including the analysis of realistic pricing models. Also, the development of a strategic business plan is key. During the growth phase, lead conversion is important. Many digital companies are using freemium models to foster initial growth. For Digital OEMs, it will usually be important to ensure initial revenue generation, e.g., for the physical parts of the offering that cannot be subsidized. Finally, when reaching maturity, converting freemium subscribers to a premium subscription will be key. Up- and cross-selling can generate additional revenues (Fig. 11.5).

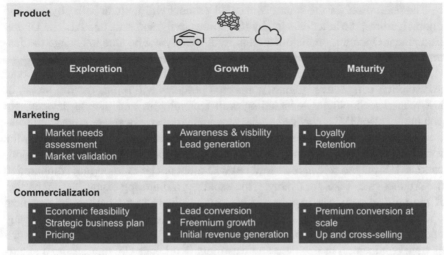

Fig. 11.5 Holistic product, marketing and commercialization strategy

11.4 Ensure Product/Market Fit (or Solution/Internal Demand Fit)

A key prerequisite for successfully establishing a scalable high-tech business is to constantly focus on product/market fit. This means that the product – including the user experience (UX), the feature set and the value proposition – must meet the undeserved needs of the target customers. Since the target customers are likely to also change over time, the organization must be able to react to the changing needs.

So who are the target customers? Digital OEMs operating in a B2C market will usually address needs such as convenience and offering cool, new features. For those in B2B markets, customers are more likely looking for efficiency improvements and cost reductions. The Digital Equipment Operator, on the other hand, will focus on operations effectiveness (OEE).

And what about the underserved needs? In terms of the target customers, the Digital OEM will usually address either a B2C or a B2B market, while the solutions for the Digital Equipment Operator will more often address the internal business units responsible for the physical assets.

How can this be met with a matching offering? The value proposition is defined by key capabilities. In "*How Smart, Connected Products Are Transforming Competition*" [11], Michael Porter and Jim Heppelmann describe four key capabilities of smart, connected products: monitoring, control, optimization, and autonomy. In the context of our discussion, the new products and services provided by the Digital OEM will probably most benefit from control and autonomy, while the Digital Equipment Operator will utilize the monitoring and optimization capabilities for his solution.

The feature set for the smart, connected product will include both physical and digital features, the latter enabled by IoT connectivity, software and AI. The Digital Equipment Operator, on the other hand, will usually not be able to change the features of the existing physical assets, so the focus here is on digital features.

From a UX point of view, we will again have to differentiate between the product and solution perspective – smart, connected products will utilize the full breath of UX-related technologies – including web technologies and mobile devices, but potentially also HMI embedded on the physical product. For smart, connected solutions developed with a focus on improving the operations of existing assets, the focus will often be more on utilizing basic web technologies for intranet-type applications. UX plays a role here as well but will often not be as important compared to the product side. Take, for example, the escalator monitoring example from earlier. If the railway company is only making this available to a small set of technical operators, a simple UX will be sufficient. This would obviously be different for any operations support apps that are made available to a wider internal audience, such as train conductors. In this case an investment in a better UX (e.g., using a smartphone app) will be justified (Fig. 11.6).

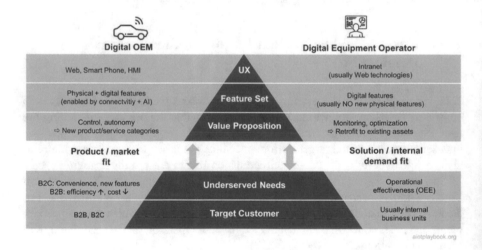

Fig. 11.6 AIoT product/solution – market fit

11.5 Ensure Efficient Exploration

Ensuring efficient exploration of new, AIoT-enabled opportunities is key for initiating scalable businesses. This is usually less of a problem for startups but something that larger companies and incumbents can struggle with.

The first challenge is the selection of suitable use cases. These should on clearly identified application areas and customer benefits. Properly evaluating the use cases with respect to their potential to scale is important as well.

The next point is "*freedom to experiment*": Especially in the early phases of exploration and technical feasibility assessment, the team should not be slowed down by corporate rules, standards and other dependencies. Instead, the focus should be on value creation and validation.

Infrastructure is important. Reuse of corporate infrastructure can make sense, e.g., in areas such as user management, billing, invoicing, and technical AIoT infrastructure. However, the exploration team should not be forced by corporate mandates to play guinea pigs for new, immature infrastructure.

Cost are essential as well. One should not apply typical enterprise cost structures during the early exploration phase. However, the team must prepare early for the target cost model further downstream.

Finally, "*enterprisification*" plays an important role, especially in larger organizations. At some point in time, the team will have to adapt to corporate governance rules, standardization, and integration into the enterprise application landscape with all its governance mechanisms. The point for full integration must be chosen carefully. Doing it too early can potentially kill the endeavor simply by adding too much overhead to a yet unproven and probably unprofitable early-stage idea. Doing it too late can also prove to be difficult because certain standards and compliance levels are necessary even before onboarding the first customers at scale. However, the enforcement of corporate rules can be a real distraction from business goals. Consequently, this should be treated like a consolidation project (Fig. 11.7).

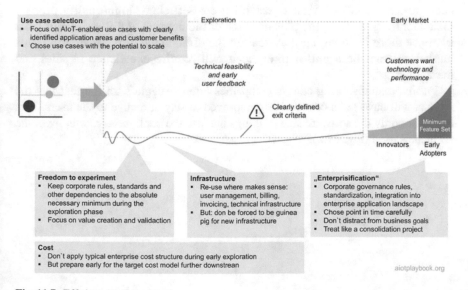

Fig. 11.7 Efficient AIoT exploration

11.6   Understand How Best to Cross the AIoT Chasm

The business book classic *"Crossing the Chasm"* by Geoffrey Moore [12] describes the challenges of marketing high-tech products, especially focusing on the chasm between the early adopters of a product and the mainstream early majority. This concept is especially important from the point of view of the Digital OEM.

Throughout the life cycle of the product, he will face a number of different challenges, some of which will be very specific to AIoT. For example, in the early stage when addressing innovators, a challenge is to actually create a small series of physical products that appeal to innovators in combination with digital features. Especially if AI is heavily used, this can be challenging in the early phase of the product life-cycle because in this phase, few reference data will be available for training the AI models. For asset intelligence enabled by AI, this will probably mean that simulation and other techniques will have to be applied. For any swarm intelligence required by the product, this will be even more challenging because the "swarm" of products in the field actually creating data will be relatively small.

When moving on to the next phase, serving the early adopters, an AIoT-enabled product will have to make difficult decisions about the MVP or baseline of the physical side of the product because this will be very difficult to change after the Start of Production. Another key point will be a strong UX to appeal to early adopters: something that start-ups tend to be better up than incumbents.

Finally, when crossing the chasm to the early majority and realizing significant growth, it will be vital to establish cost-efficient, high-quality product manufacturing. Scaling the physical side of the product at this point will most likely be more challenging than scaling the digital side of it. In addition, this market will not be a pull-market, so it will require excellence in sales and marketing.

Finally, manufacturing-centric companies often struggle with the fact that the product will have to be continuously improved to stay attractive to the users. This means not only the software side of things but also the continuous retraining of the AI models used (Fig. 11.8).

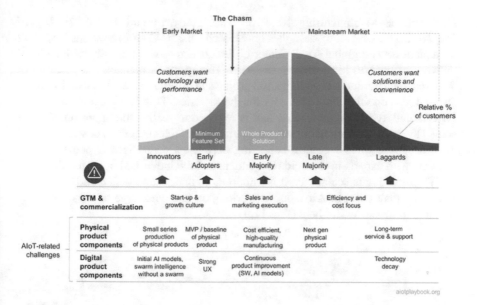

Fig. 11.8 Crossing the AIoT chasm

Gabriel Wetzel, CEO of Robert Bosch Smart Home: *"A key challenge are the often very high expectations, which don't anticipate the 'trough of disillusionment' which you usually have to cross before you will see new business at scale. You have to make sure to get through this, and not lose management support on the way."*

11.7 Understand Implications of AIoT Short Tail vs. Long Tail

A good way of looking at the scalability of the opportunities presented by AIoT is by categorizing them in the short tail vs. the long tail of AIoT: the AIoT short tail includes a relatively small number of opportunities with a high impact and thus high potential for scaling them. This usually means a high level of productization and a strong Go-to-Market focus, which requires a Digital OEM organization. The AIoT long tail, on the other hand, represents a large number of opportunities where each individual opportunity is relatively small. However, together these long tailed opportunities also represent a very significant business opportunity, provided an organization is able to harvest these smaller opportunities in an efficient way. This usually requires a "harvesting" type of organization (for internal opportunities) or a platform approach, as described earlier.

A good example for an organization that is focusing on harvesting a large number of small opportunities presented by AIoT is described in the "AIoT in

High-Volume Manufacturing Network" case study in Part IV of the *Digital Playbook*. This case study describes how Bosch Chassis Control Systems have built up a platform and global AIoT Center of Excellence to work closely with a global network of over 20 high-volume factories. This group is managing a portfolio of AIoT-enabled production optimization projects in different areas, but usually with a strong focus on OEE improvements for the factories. This is a great example of a "harvesting" type of organization that is required to realize the opportunities presented by the AIoT long tail in such an environment. Executing this at scale for over 30 factories requires a careful balancing between a centralized expert team and working with experts in the field who understand the individual opportunities.

When looking at scalability, it is important to understand which end – the short tail or the long tail – of AIoT one is working on and what type of organization is required to be successful here (Fig. 11.9).

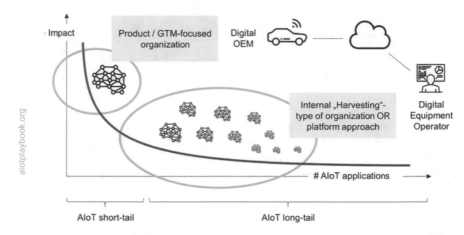

Fig. 11.9 The long tail of AIoT

Gabriel Wetzel, CEO of Robert Bosch Smart Home: *"The short-tail opportunities will often be addressed by other market players as well. This means that investment size and time-to-market are absolutely critical. The long tail requires many custom solutions. You should not underestimate the required resources, the domain-specific skills and the market access. Not all of these can be easily scaled-up. Of course this can be addressed by a top-in-class partner management: but don't forget to budget for it!"*

11.8 Ensure Organizational Scalability

Ensuring organizational scalability is another key success factor for smart, connected products. How can an organization successfully grow and evolve alongside the product as it matures from idea to large-scale business? DevOps mandates that an IT organization combine development and operations from the beginning, iterating together continuously through the build and improvement cycles. However, in an organization that must combine IT development and operations capabilities with physical product engineering and manufacturing capabilities, this will not be as straightforward.

Dattatri Salagame is the CEO of Bosch Engineering and Business Solutions. In the following, he will discuss the issues related to scaling up and evolving an organization for smart, connected products.

Dirk Slama: *What is your take on the organization we need to build and sell smart, connected products?*

Dattatri Salagame: *Since AIoT is a relatively new space, organizations are finding their feet to unlock potential at scale. A typical AIoT product would need multiple players to come together conceive, develop and launch a connected product. As data twins are the backbone of the connected products, the diversity and complexity of the technology stack demands players with deep tech domain, cloud platforms, and connectivity to come together to orchestrate the end product or service. Most AIoT projects go through different product phases without transition awareness of shifting to different phases in terms of Product Lifecycle Management (PLM) shift and capability needs. A connected product engineering organization needs to evolve with the product. This transition has to be managed while the business owner is able to do their experimentation, validation in the market for scalability.*

Dirk: *What are the required organizational capabilities during these different phases?*

Dattatri: *In the beginning you need a gang of hackers who can quickly hack a solution in a high-fidelity experimentation mode. I call them a gang of hackers because in the MVP (Minimal Viable Product) stage, you hack the solution, you are not really worried about the reliability of the product. You are worried about the feasibility of the product. Once you have confirmed the feasibility and you have received positive customer feedback, you need to transition into a lot more rigorous, disciplined product engineering process. Multidisciplinary product engineering has become a key competency. You will require competencies encompassing mechanical, electrical, and electronics engineering, power management, communication, coupled with data science, AI and security. It is very important that there are team members who understand data and mathematical modeling of the data to mimic the physics and chemistry of the product, to create AI models and to be able to mature the product in layers.*

The product matures in layers, which is important. The electronics layer, the communication layer, the network service provider, then the data, then the AI and then the reliability. So these layers mature at a different velocity, at different points of time, so you need a team which is a lot more multi-disciplined and engineering rigor in this phase. Finally, when we release the product, and we have crossed the initial validation of the scale, then you need an organization to support the product introduction. This is a game of having a good ecosystem to be able to manage the scale and to provide high-speed DevOps as the backbone of digital services. Therefore, you need these three flavors of the team during the course of the project:

- *The Minimum Viable Concept/Product (MVC/MVP) – feasibility validation*
- *New Product Engineering (NPE) – Reliability validation*
- *New Product Introduction (NPI) – Scale and ecosystem*

Dirk: *Any recommendation on how to organize this?*

Dattatri: *Everybody is talking about digital transformation, and this is it: we need to transform existing, traditional, heavy engineering and manufacturing-oriented organizations, so that they play together with the more agile AI and software organizations to support smart, connected products. It is important to be transition aware through the phases of the product life cycle in terms of PLM and capability shifts. Our experience has been to adopt a multi-speed model to navigate through these phases, with clear "Transition Awareness" to operate in right gears. To manage capability shifts, one needs to operate in a multi-threaded model covering classic product engineering to technology (AIoT) fusion. Otherwise, there is a risk falling through between the transitions, which we call the "valleys of death". If you look at connected products, seven or eight out of ten products don't actually pass in flying colors. Therefore, organizational agility and the ability to transform the organization along the way is an important part of the ability to "cross the chasm", as you have introduced it earlier (Fig. 11.10).*

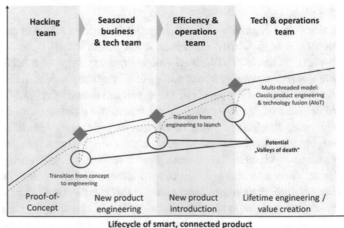

Fig. 11.10 AIoT organizational evolution

11.9   Deal with Repeatability, Capacity and Marginal Costs

Digital businesses are seen as potentially highly scalable because their digital offerings are highly standardized and easily repeatable at very low extra cost. Physical products, on the other hand, can be much harder to scale, because scale effects in manufacturing often only apply when talking about extremely high production numbers. Even in this case, the marginal costs will not be reduced to a level as we are seeing this in the case of digital businesses.

For the Digital OEM, this means that their focus usually needs to be on creating highly standardized physical products, because any increase in variants and added complexity can potentially have a negative impact on scalability. Ideally, differentiation through product variations should be mainly focused on the software/AI side. An interesting example in this context is the *Seat Heating-on-demand* case, which is introduced in the product operations section: instead of having cars manufactured with individual seat heating configurations, all cars come with the same physical equipment and the configuration is done later on-demand by the customer. Of course this type of business case requires careful calculation of the marginal production costs vs. the downstream revenue opportunities over the life-cycle of the car.

For the Digital Equipment Operator, the topic of repeatability and capacity is also important. This links closely back to the long-tail discussion from earlier on: if the benefits of the individual AIoT-enabled solutions are only relatively small in comparison, then ensuring repeatability on some level is key. In the *"Predictive maintenance for hydraulic systems"* case study, Bosch Rexroth used AIoT to enable predictive maintenance for hydraulic components. However, since each customer installation uses the hydraulics components in a different way, AI algorithms have to be adapted individually for the customer. Bosch Rexroth has established a service offering that maximizes repeatability by standardizing the sensor packs and establishing a standardized process for the customization of AI for individual customers. In this way, the predictive maintenance service offering is competitive, despite its positioning on the AIoT long tail.

Part III
Business Execution

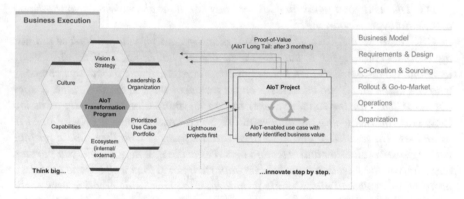

Fig. 1 Business Execution

The Business Execution part of the Digital Playbook provides guidance on how to actually implement the AIoT-enabled business strategy. In the following, C K Vishwakarma (CEO of AllThingsConnected & co-lead of the AIoT User Group in Singapore) shares some of his thoughts on this.

Dirk: *The AIoT transformation program. Is that for real or is it a synonym for a generic digital transformation program?*

C K Vishwakarma: *It is indeed real because if we tried to understand slightly more specific, that is the definition of AIoT itself. As we know, AIoT is the combination of AI and IoT. This means that AIoT is applied to certain very specific use cases, related to physical assets or products. Digital transformation is very broad, and within your digital transformation initiative (which may have hundreds of such initiatives), we can identify those with a direct relationship to physical assets and products. Those I would say will be driven by the AIoT initiative. So the organization have to think of the bigger objectives, and then there are complex objectives within which will be categorized as AIoT in my view.*

DS: *What are the first steps for setting up an AIoT transformation program?*

CK: *When we are initiating the AIoT program, we need to think about what the business objective is, what are required process transformations or operational challenges, etc. It is not just because everybody else is doing it, so we should do it as well. The first step is obviously, what are the main points? What are the challenges that you are trying to address? What are the use cases in those initiatives? Then we will look at what kind of technology you need to put in place and finally, how to build a team together and implement the use cases. Understand what is in there for us as an organization, because the way we look at the way we design, manufacture, assemble, distribute, and operate products and services are going to change. So where are the opportunities specifically, what process gap problems do we have, and how can we convert that into opportunities. And then we can look into evaluating the right technology within the whole AIoT spectrum. Finally, how do we implement the required AIoT products and solutions, with an in-house team or with a development partners?*

DS: *That is a good point. In AIoT, nobody can do it alone. What's your take on co-creation versus sourcing?*

CK: *Yes, absolutely. Nobody can do it alone because if you just look at not just the business challenge of it, but also the technical challenges, AIoT requires many technologies to come together. I have not come across any one technology company who can claim that it can address the entire technical stack at the same time. It does not make sense for the solution seeker to do everything alone. It usually takes too long, and you risk losing the opportunities. So co-creation part is critical in AIoT initiatives. The challenge that most organization faces is who they should partner with. Partner evaluation and selection becomes critical. Where should you partner, and what are the selection criteria? It is important not to look well beyond the initial proof-of-concepts and pilots. While it important to show value creation very early, most AIoT initiatives are also very long-term oriented. Therefore, identifying the right areas for partnering and co-creation is important. And of course partnering vs buying. So we are going from make-vs-buy to make-vs-buy-vs-partner.*

Dirk: *Thank you, CK!*

Chapter 12
Business Model Design

Dirk Slama

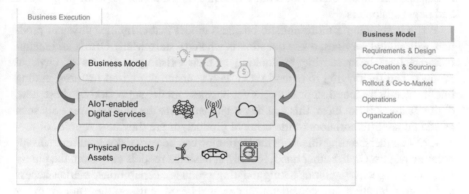

Fig. 12.1 BMI

The development and validation of a (more or less) detailed business model is usually the first step in the journey towards developing a new smart, connected product or solution. The business model will usually play a central role in product development, while in solution development, it will be most likely more basic. This chapter looks at tools and best practices for developing AIoT-specific business models, with a focus on the product perspective (Fig. 12.1).

D. Slama (✉)
Ferdinand Steinbeis Institute, Berlin, Germany
e-mail: dirk.slama@bosch.com

© The Author(s) 2023
D. Slama et al. (eds.), *The Digital Playbook*,
https://doi.org/10.1007/978-3-030-88221-1_12

12.1 AIoT-Enabled Business Models

For a product or service-oriented organization, the business model usually describes how the organization creates, delivers, and captures value [13]. For example, the St. Gallen Business Model Navigator [14] defines four dimensions for a business model ("What, how, who and value"): What do you offer to the customer? How is the value proposition created? Who is your target customer (segment)? How is revenue created?

For a more operations-oriented organization looking to introduce AIoT-enabled solutions, the business model will often be developed around the OEE formula introduced in the discussion about the Digital Equipment Operator: How will the investment in a new, AIoT-enabled solution benefit asset availability, performance rate or quality rate. The focus here will usually be more on the business case (ROI, OEE), and less on the other aspects of the business model such as value proposition and target customers.

When developing a more holistic business model that combines physical products with digital services, a key question is where to start. Many OEMs and manufacturers have been starting by looking at their existing portfolios of physical products and then trying to extend them with connectivity-based features, adding intelligence in a second step. If the target business model supported by these new, digital features is not clear, this can be problematic: Are the new features only seen as additional differentiators of the original product, or are they new sources of revenue? Not understanding this from the beginning can be very risky, leading to disappointing results. On the other hand, looking at business models only from the purely strategic perspective without taking existing products, capabilities, market access and brand reputation into consideration can also be difficult. So the truth probably lies somewhere in the middle.

Christian Renz, Global Head of IoT and Digitalization at Bosch, has made the following experience in this area: *Successful manufacturers of physical products typically have a great understanding of their products' domain: They understand usage of their product, value creation processes they are part of, the competitive landscape, purchasing behavior and so on. However, they view their domain through the lens of their products, which means they tend to not perceive value creation that their products are not part of. To successfully incubate new AIoT business models, product manufacturers need to build up competencies in service incubation and design. Coming from a hypothesis of value created for customers, dedicated teams quickly iterate through a series of "minimum viable products", proving or disproving the value hypothesis. These cycles are much faster than typical product engineering cycles. The initial value hypothesis should be purely derived top down from concrete customer pain points in the domain, rather than bottom up from augmenting physical products with connectivity, allowing even for the freedom to come up with business models that could potentially disrupt existing business* (Fig. 12.2).

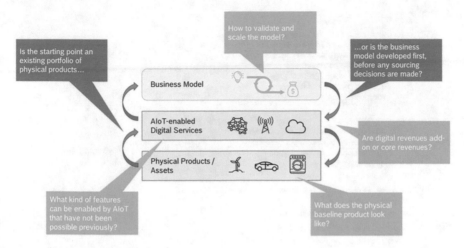

Fig. 12.2 AIoT-business model development – considerations

What is important to understand is that business models usually evolve over time. In the agile and lean world, the assumption is that business model innovation is an iterative process. Many Internet-based business models are constantly re-evaluating and adapting their business models, utilizing the flexibility of the cloud and DevOps to do so. However, for business models based on physical assets, this is typically not as easy: Design and manufacturing of physical assets has much longer lead times. Once the assets are manufactured, sold and deployed in the field, any alteration of their physical configuration becomes very difficult if not impossible. Smart, connected products provide an opportunity to address this issue, at least to a certain extent, for example through dynamic configuration of digital features or, in some cases, even the enabling of physical features on demand. The following will discuss typical business model patterns enabled by AI and IoT in more detail.

12.1.1   AI Business Model Patterns

The area of business model patterns based on AI in the context of the IoT is not (yet) widely researched. Figure 12.3 describes the most common patterns from the AIoT perspective.

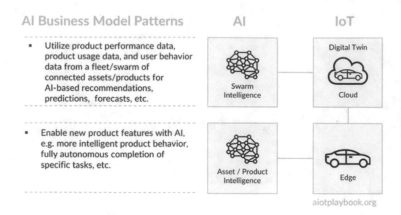

Fig. 12.3 AI business models patterns

12.1.2　IoT Business Model Patterns

The area of business model patterns for the Internet of Things is well researched and documented. For example, the St. Gallen Business Model Navigator [14] defines a number of patterns summarized in Fig. 12.4. These patterns are generally based on the assumption that they combine physical assets with digital services.

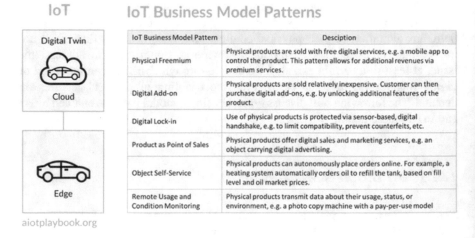

Fig. 12.4 IoT business model patterns

A great example of a 'Digital Add-on' is BMW's announcement to make seat heating available on demand. Two factors make this interesting:

- Physically producing many different, custom configured variants of a car could be nearly as expensive as producing a single, mass-manufactured variant
- Being able to upsell this feature to customers especially in winter could significantly increase the total number of seat heating options sold in total

12.2   Ignite AIoT Business Model Templates

The following introduces a set of templates for AIoT business models. As far as possible, these templates reuse existing, well-established business model templates, adding AI and IoT perspectives to them. These templates should be seen as guidance and can be adapted in a flexible way to best fit the needs of your individual AIoT business model (Fig. 12.5).

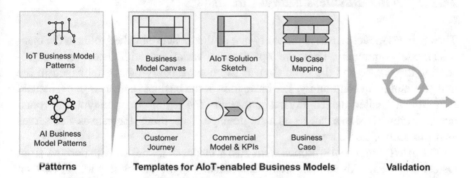

Fig. 12.5 Ignite AIoT business model templates

12.2.1   The Smart Kitchen Example

The following discussion will be based on the smart kitchen example, which is shown in Fig. 12.6. The complete Smart Kitchen Business Model has been documented in Miro. It can be accessed HERE, in case you can't read some of the details in the following diagrams.

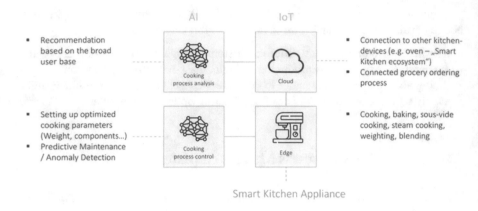

Fig. 12.6 Example: smart kitchen

12.2.2 AIoT Business Model Canvas

The business model canvas is probably one of the most established tools in the business model community. There are a plethora of variations, with Osterwalder representing the classic and the Lean Canvas the one probably most established in the agile development community. The basic idea of the business canvas is that – instead of writing a detailed and lengthy business plan – the key information typically found here is summarized in a canvas on a single page. Sometimes, the canvas also serves as the executive summary.

The AIoT Framework proposes an AIoT business model canvas derived from Osterwalder, but adds another area to specifically highlight the impact of AIoT on other elements, including value proposition, customer relationships, channels, key resources, key activities, cost and revenue structure (Fig. 12.7).

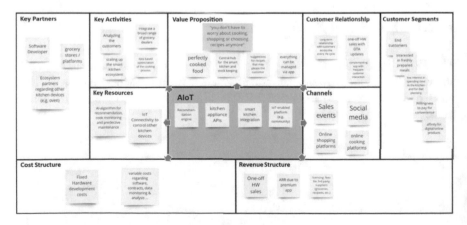

Fig. 12.7 AIoT business model canvas

12.2.3   AIoT Solution Sketch

The first template is the so-called *AIoT Solution Sketch*. The idea is to provide a very simple canvas which helps visualize the key functional elements of your solution, mapped to either the field (including EDGE functionality) or the back-end (e.g., in the cloud). This simple yet expressive format is especially useful for reviewing and discussing the intended functional scope with management stake-holders (Fig. 12.8).

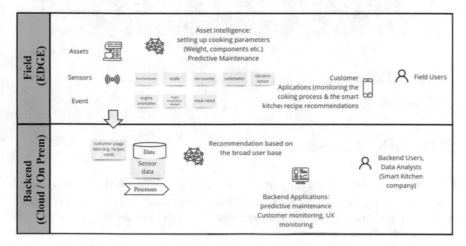

Fig. 12.8 AIoT solution sketch

12.2.4   AIoT Use Case Mapping

AIoT Use Case Mapping can be used to clarify how far one of the typical AIoT Use Cases can best be supported by utilizing AI and IoT together. An example is given here. In the example of the kitchen appliance, almost all generic AIoT use cases are relevant: AIoT will be used to improve the design of the kitchen appliance, data will be used for sales support, overall product performance improvements are in scope, as well as predictive maintenance. Finally, digital services such as recipe recommendations will play an important role (Fig. 12.9).

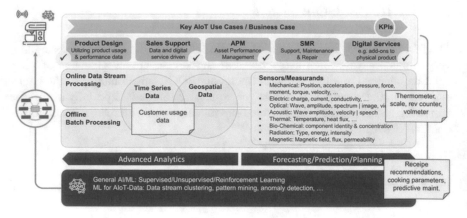

Fig. 12.9 AI value proposition – smart kitchen example

12.2.5 AIoT Customer Journey Map

Customer (or User) Journey Maps are a common User Experience (UX) tool. There are many shapes, sizes, and formats available. The general idea of a journey map is to help understand and visualize the process that a person goes through to accomplish a specific goal.

The Digital Playbook proposes a format for a customer journey map that has the key user interactions with the asset at the top, e.g., asset purchasing, asset activation, asset usage and service incidents. Depending on the complexity, each of these steps could be detailed in a map on its own. Below this, the template provides space for the following:

- **Touchpoints**: What touchpoints is the customer actually using to interact with the solution or the asset?
- **Doing**: What is the customer actually doing?
- **Thinking/Feeling**: This covers the emotional side of the journey
- **Opportunities**: What opportunities from a business model point of view can be found here?
- **Key AIoT Features**: What features/capabilities from an AIoT point of view are utilized here?

Note that this template focuses more on the high-level journey, including business model aspects. A more detailed, UX-focused version of this is introduced later in Product/Solution Design (Fig. 12.10).

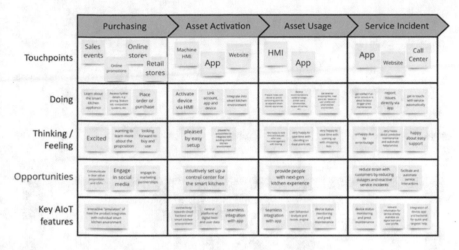

Fig. 12.10 AIoT customer journey

12.2.6 *Commercial Model*

The commercial model has to address the question of how the product or solution is generating revenues at the end of the day. The model must bring together the offering and the target customer.

The offerings must be broken down as follows:

- Unique value proposition: potentially for different customer segments
- Sellable features: identify all elements of the offering that eventually generate revenue, e.g., upfront revenues for the physical asset, subscription revenues for digital premium services
- Pricing: all sellable features must be included in the pricing model

The target customer must be well understood, including:

- Industry/domain: this will look different for B2C vs B2B offerings but should be addressed, e.g., via a market segment analysis
- Profile: again, must be looked at individually, e.g. B2B buying-center vs B2C persons
- Buying process: how is the customer – as a private person or an enterprise – buying this? What formal conditions have to be met?

Finally, the question is how to get to the customer:

- Sales approach: traditional Solution Sales and Key Account Management, web-based sales, in-app sales, etc.
- Monetization: how to turn non-revenue-generating items into revenue, e.g., by getting customers to upgrade from digital fremium to premium services (Fig. 12.11)

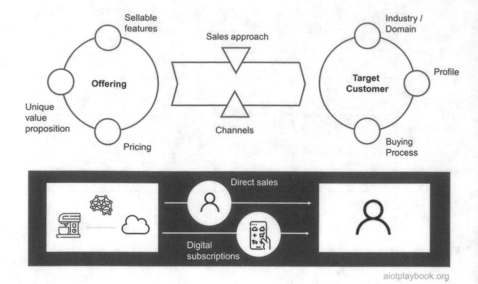

Fig. 12.11 Commercial model

12.2.7 KPIs

It is usually advisable to include a set of Key Performance Indicators (KPIs) in the Business Model. KPIs are measurable values used by organizations to keep track of and determine progress on specific business objectives. A good method for defining KPIs is the SMART method. SMART means that KPIs should be specific, measurable, attainable, relevant, and time-sensitive. Example KPIs for an AIoT product are described in more detail in the Product Design section.

KPIs are a good way of guiding the execution team and evaluating their progress against a previously defined set of goals in the business model. Closely related to this are Objectives and Key Results (OKRs). While KPIs are business metrics that reflect performance, OKR is a goal-setting method which can be used as a project steering mechanism. However, this would usually not be part of the business model.

KPIs for the kitchen appliance example could include:

- Number of kitchen appliances sold
- Average subscription revenue per customer
- Monthly recurring revenue
- Customer Lifetime Value
- Customer acquisition cost

These KPIs are assuming an established business. In the early phase of business model validation, different KPIs should be applied – for example, KPIs related to UX.

12.2.8   AIoT Business Case

Another key element of the business model is the business case, including the financial perspective on costs and revenues, as well as strategic contributions.

Direct ROI

The direct ROI for an AIoT solution must typically take into consideration asset-related and service-related costs and revenues. On the cost side, the differentiation between capital expenditures and operational expenditures (including unit and operations costs). On the revenue side, the business case must differentiate between upfront revenues and recurring/subscription revenues. The Digital Playbook proposes combining these perspectives in the template shown in Fig. 12.12.

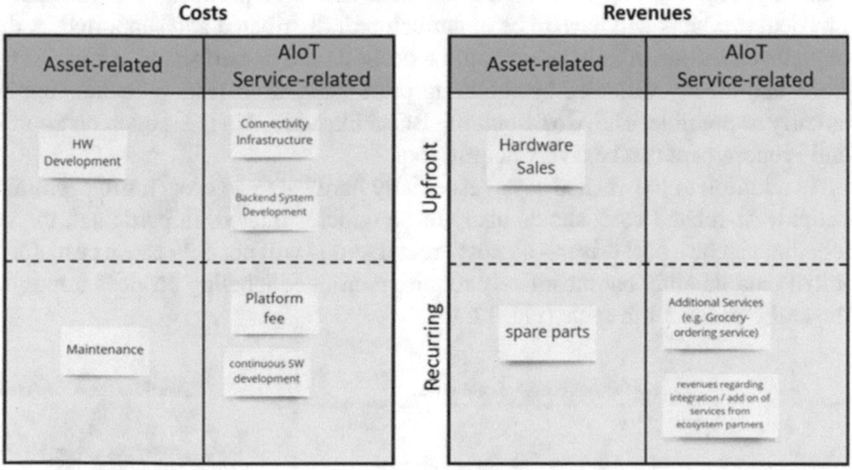

Fig. 12.12 AIoT business case

Please note that business case development and ROI calculation usually also require some kind of quantitative planning, including projections for numbers of units sold, customer adoption of digital features, and so on. A detailed example is given in the Product Design section.

Strategic Contributions

In addition to the direct ROI of the investment, many AIoT solutions also provide strategic contributions to a higher-level business case. Take, for example, the eCall feature of a car. This potentially AIoT-enabled device in vehicles will automatically call a local emergency service in the event of a serious road accident. Airbag deployment and impact sensor information, as well as GPS coordinates, will be sent along as well. The question is as follows: Does this feature require a dedicated ROI calculation, or is this simply the fulfillment of a regulatory requirement? Since eCall is now a requirement in the EU, for example, it is unlikely that this can be sold as an add-on with extra revenue. So it must be seen as a strategic contribution to the car.

12.2.9   AIoT Business Case Validation

Validating the AIoT Business Case in the early stages as much as possible will save you from costly surprises further down your AIoT journey. The business case validation should include both sides, costs and revenues.

Validating assumptions made about revenue in the business model is of course tricky. Usually, a good way forward is interviews with potential customers to validate not only their willingness to purchase the intended products and services but also their price sensitivity.

Furthermore, one should also not underestimate the importance of validating the cost side of the business model. This is especially important for an AIoT-enabled business: While virtual, cloud-based business can scale very well on the cost side, with any business that is involving physical assets or products, this is different. Physical products will have to be manufactured, distributed and supported. A thorough investigation of unit costs/marginal costs should be performed as early as possible, and ideally validated by obtaining price indications from potential suppliers as early as possible. The AIoT Sourcing BOM introduced in the section on Sourcing and Procurement can be a very helpful tool.

In addition to IoT-related costs (especially hardware and costs for telecommunication), AI-related costs should also not be underestimated. In particular, the data labeling can be a cost driver – do not forget that this will not only cause costs for the initial data labeling but most likely require continued labeling services throughout the entire product life cycle (Fig. 12.13).

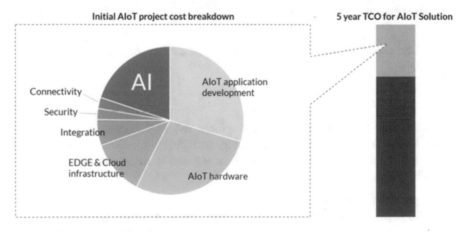

Fig. 12.13 AIoT cost estimation

In general, IT-centric business cases have a tendency to focus more on the initial costs, and not the Total Cost of Ownership (TCO). Over a five-year lifespan, initial development costs will most likely be only 20% of the TCO [15].

12.3   Proof of Concept

Most investors require some kind of proof along the way, which provides evidence for the feasibility of the investment proposal (this applies both to corporate investors and private equity investors). AIoT-based solutions are not different from that perspective. However, it can sometimes be much more difficult and expensive to run a *Proof-of-Concept (PoC)* for an AIoT solution: Today it is usually very easy to create a lightweight and affordable PoC for a pure software project (e.g., using simulation or mock-ups). However, as soon as hardware development and/or asset customization is involved, this can become much harder, depending on the hardware and asset categories.

Consequently, the following should be clearly defined for any AIoT-related PoC:

- Duration & effort
- Scope
- Resources
- Success criteria

12.4   Investment Decision

In today's agile and digital world, most investment decisions are staged, meaning that partial investment commitments are made based on the achievement of certain milestones. However, there is usually a point in time for any innovation project where it transitions from the *exploratory phase* toward the *scaling phase* with much higher budgets. Each organisation is typically follows its own, established investment criteria. For the project manager, it is often important to keep in mind that these criteria are usually a mixture of hard, ROI-based criteria, as well as the strategic perspective. This is why the business model should address both perspectives, as stated above (Fig. 12.14).

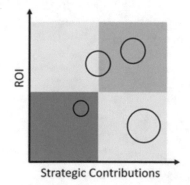

Fig. 12.14 AIoT investment decision

Chapter 13
Product/Solution Design

Dirk Slama

Fig. 13.1 0.2 Design.png

What is the next step after agreeing on the first iteration of the business model and securing funding for execution? With the traditional waterfall approach, the answer is relatively clear: documentation of requirements with a high level of detail and accuracy, serving as the stable foundation for planning, design and execution. However, in the software world, we have learned that getting stable, long-term requirements is often difficult. Consequently, traditional requirements management often has a bad name in the agile community. Instead, agile best practices focus on the backlog as the central means of managing requirements and work items. The goal is to capture the high-level, long-term vision on a more abstract level (e.g., via so-called epics and themes) and then provide a detailed and precise work definition only for the next upcoming sprint (via the sprint backlog – created for each sprint;

D. Slama (✉)
Ferdinand Steinbeis Institute, Berlin, Germany
e-mail: dirk.slama@bosch.com

© The Author(s) 2023
D. Slama et al. (eds.), *The Digital Playbook*,
https://doi.org/10.1007/978-3-030-88221-1_13

typical sprint duration is 3–4 weeks). In this way, the agile approach ensures that no waste is created by investing in detailed, long-term requirements that are likely to change over time anyway (Fig. 13.1).

Most AIoT projects or product developments will need to combine both perspectives. For the standard software part – and probably also the AI part – agile planning will work well. For those parts involving hardware and telecommunications infrastructure – as well as any part with complex dependencies – a more planning-centric approach will most likely be required. Enterprise sourcing processes will add their part to limit a purely agile approach.

Nevertheless, a good starting point will be the agile best practices in this area, which we will discuss first. Next, we will look at how to derive an AIoT system design from all of this. This will be completed by a discussion of the entire cycle from requirements and design to implementation and validation. Finally, we will discuss the dependencies between AIoT system design and co-creation/sourcing.

13.1 From Business Model to Implementation

The Digital Playbook proposes an approach that combines the business model design patterns outlined in Part III with Agile Story Maps, as well as an AIoT-specific approach for product/solution design. This is shown in fig. 13.2. The starting point are the elements of the business model design, which can be adapted over time based on market feedback. The product/solution design provides a lightweight yet holistic view of the system design, from the business view down to the implementation view. Finally, the Agile Story Map provides the high-level breakdown of all work items. For each sprint, a dedicated sprint backlog is derived from the story map, containing the prioritized and agreed-upon work items for the upcoming sprint (Fig. 13.2).

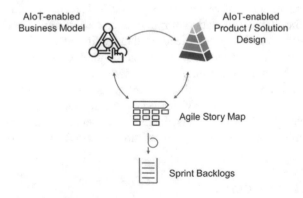

Fig. 13.2 From business model to implementation

13.2   The Agile Approach

The agile equivalent to requirements management is the backlog. The product backlog is a list of all the work items required to build and improve the product. It represents the single source of work definitions accepted by the scrum teams. For each
upcoming sprint and each sprint team, a sprint backlog is created, which defines the
work to be done by the team in the next sprint. Product backlog items can have different granularities. Sprint backlog items must be implementable in a single sprint.
Story maps often serve as the visualization of the product backlog, as will be
described in the following.

13.2.1   Story Maps

Story maps are a useful tool to manage the high-level requirements and structure of
a product. Depending on the school of thought, they are either described as a visualization of a product backlog or as a customer journey-centric way of structuring
the body of work on the highest level. Especially in the early stages of product
creation, story mapping is used as a technique for product discovery, helping to
outline the structure of a new product (or a complex, new feature for an existing
product). To achieve a higher level of abstraction and orientation than a linear backlog, story maps typically arrange lower-level features in higher-level, functional groups.

Typical units of work in story maps include so-called themes, epics, features and
user stories. The number of levels in the hierarchy of the body of work depends on
the complexity of the product/solution. For the purpose of simplicity, the Digital
Playbook focuses on epics, features and user stories:

- Epics are a high-level body of work, typically representing 2–6 months in duration. An epic can span multiple releases and more than one team. They are often
  aligned with senior management. Epics contain features.
- Features describe a specific functionality of the product. They are smaller than
  epics and typically contained within a specific release and assigned to a specific
  team (see ⇒ feature team). They are typically managed by product owners.
  Features contain user stories.
- User stories are the smallest definition of an increment, usually less than a week.
  They are bound to a specific sprint.

Depending on the layout used, the story map can imply a certain order at the top
level, either in terms of the logical sequence in the customer journey or the order on
which the elements are worked (Fig. 13.3).

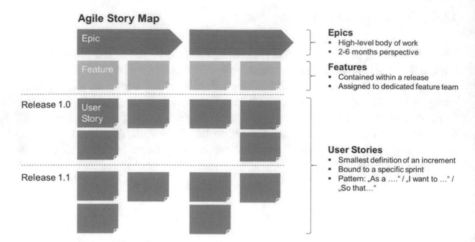

Fig. 13.3 Story map overview

13.2.2   Example: AIoT Story Map & User Stories

The example in Fig. 13.4 shows a more detailed structure of a story map. The top level contains epics. Returning to the smart kitchen appliance as an AIoT example from earlier, epics at this level could include *"cooking"*, *"baking"*, and *"recipe recommendations"*. On the level below, features are shown. A feature of the smart kitchen appliance could be a *"predefined baking program"*. Below this, user stories are shown. Use stories usually follow a specific pattern, as shown in the figure below (*"As a..."*). User stories should also include acceptance criteria. Depending on the layout of the story map chosen, use stories can also be grouped into releases.

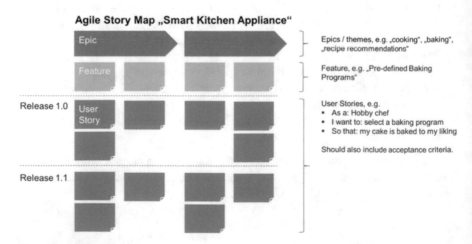

Fig. 13.4 Story map details

13.3 Non-Functional Requirements

Story maps, including epics, features and user stories often focus on the functional aspects of the system. In addition, most AIoT solutions or products will usually also have some nonfunctional requirements (NFRs). NFRs usually define attributes such as availability, performance, reliability, scalability, security, maintainability, and usability. For a distributed, AIoT-enabled system, NFRs might have to be broken down to specific areas, e.g., edge vs. cloud or specific functional areas. For example, an autonomous robot might have different NFRs for functional safety-relevant vs. non-relevant areas. This is important to keep costs and effort down to a realistic level, since functional safety development is usually much more expensive.

Finally, it should be noted that nonfunctional requirements for AI-enabled components are often different than those of traditional, software-enabled components. NFRs for AI components can include, for example:

- Algorithm accuracy and reliability: comparing AI output to reality
- Algorithm performance: for both online and training
- Transparency: making results explainable
- Fairness: ensuring results are fair and non-biased
- Testability: ensuring that the AI can be properly tested
- Security and privacy: related both to input and output

13.4 AIoT System Design

Some agile software projects will mainly rely on story maps and user stories, without an explicit system design. However, a more complex project may also require a certain amount of system design. Sometimes this is done in a *"Sprint 0"*, which then focuses on creating both the high-level story map as well as a corresponding system design.

Given the typical complexity of an AIoT initiative, a system design is required that helps align all stakeholders and subsystems. *The Digital Playbook* proposes a set of design viewpoints, which are introduced in the following.

13.4.1 AIoT Design Viewpoints

The Digital Playbook proposes four viewpoints to help create a consistent system design which covers all relevant aspects: business Viewpoint, UX Viewpoint, the Data/Functional Viewpoint, and Implementation Viewpoint. The initial Business Model will have a huge impact on the Business Viewpoint. The UX Viewpoint will be heavily influenced by the customer journey. Policies and regulations will have a huge impact on both the Data/Functional Viewpoint as well as the Implementation

Viewpoint. Again, technical constraints as well as skills and organization will heavily influence the implementation viewpoint. Finally, AIoT and enabling technologies will have an impact on all viewpoints. For example, car sharing, as we know it today, would not be possible without smartphones to hail rides and interact with the system. As an enabling technology, they have both enabled new business models, created a new UX, heavily influenced the system functionality, and finally the implementation (Fig. 13.5).

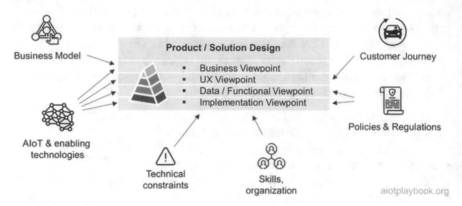

Fig. 13.5 Product/solution design – overview

13.4.2 AIoT Viewpoint Details

The Digital Playbook provides a set of templates for each of the four viewpoints. They are discussed in detail in the technology execution section on product/solution design. The Business Viewpoint starts with input from the Business Model, and then adds key KPIs, quantitative planning and a milestone-oriented timeline. The UX viewpoint is based on customer and/or site surveys and focuses on personas, Human/ Machine Interaction and mockups for key user interfaces. Of course, the Data/ Functional Viewpoint includes a high-level overview of the main data domains and the component and API landscape. If the project makes use of Digital Twins as an additional structural element, this is included here as well. Finally, the AI feature map helps ensure that AI is utilized to the fullest potential. The Implementation Viewpoint provides a high-level end-to-end architecture, an asset integration architecture, a hardware architecture, and software and AI architecture. Adjacent to all of this, the Product Viewpoint includes the Story Map, the mappings to feature teams, and the sprint backlogs.

It is important that the level of detail in the different viewpoints be kept on such a level where it is useful as a high-level documentation to enable cross-team alignment and efficient stakeholder communication, without drifting into the habits of waterfall-centric, RUP-like detail models. Detailed design models should only be created on-demand and where specifically needed. Again, this is also likely to differ for the software vs. hardware parts of the system (Fig. 13.6).

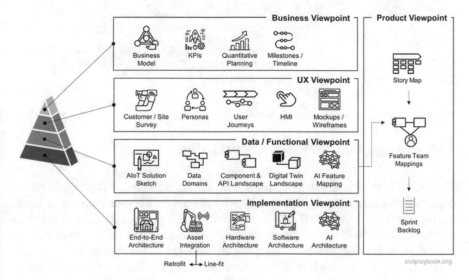

Fig. 13.6 AIoT design viewpoints – overview

13.5   From Requirements and Design to Implementation and Validation

Business models, story maps and product/solution designs capture key requirements and high-level design decisions. The teams in the DevOps organization are responsible for the implementation, testing and continuous delivery of the product increments. However, in AIoT, we cannot always assume a fully agile approach due to the aforementioned constraints. This is why we introduce the Agile V-Model, which proposes to execute individual sprints as small V-Model iterations, which take the high-level design plus the current sprint backlog as input, and then perform a *miniaturized* V-Model iteration, including Verification and Validation at the end of the sprint against the initial requirements. Finally, customer and user feedback as well as product performance data need to be incorporated back into the requirements and design perspective. This way, continuous improvement can be ensured (Fig. 13.7).

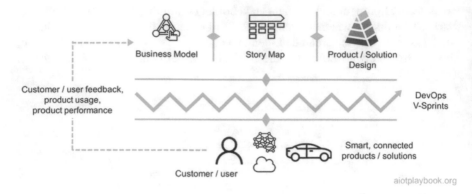

Fig. 13.7 From design to implementation and validation

13.6   Design vs. Co-creation & Sourcing

Finally, a key question that remains is who should actually do the requirements capturing and design work. This will heavily depend on the make/buy/co-creation strategy chosen. If the AIoT system is mainly developed in-house, both requirements and design will have to be created and maintained by the in-house team.

If the company decides to acquire significant parts of the system from outside suppliers or partners, high-level requirements/designs such as epics and feature definitions are likely to remain in-house. For the user stories, this could go either way, depending on the sourcing model chosen. Finally, in the case of a turnkey solution, one might even go to the extreme of only retaining the high-level requirements management in-house, and relying on external suppliers/partners for the rest.

Especially for AIoT systems with their more complex supply chains including hardware, software and AI, it is important to carefully balance this out (Fig. 13.8).

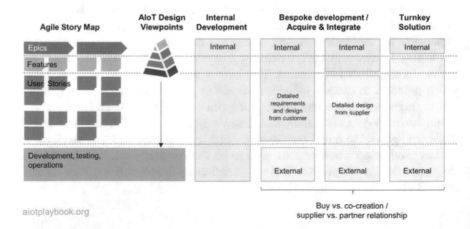

Fig. 13.8 Story map vs. sourcing perspective

Chapter 14
Co-Creation and Sourcing Intro

Dirk Slama

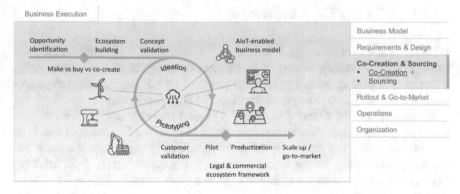

Fig. 14.1 Co-creation

AIoT-enabling a product or solution usually requires quite a number of different technical components, infrastructures and resources. Very few companies will be able to source all of this completely internally, so naturally sourcing plays an important role in most AIoT initiatives. Furthermore, there is often an opportunity to partner with other companies, e.g., because of complementary market access, brands, or product components and capabilities. Going from traditional customer/ supplier relationships towards a partnership can have many benefits but obviously many risks as well. In many cases, one will see both in the context of an AIoT initiative, e.g., traditional sourcing for commodity components and resources, and a more partner/co-creation oriented approach in other areas. This chapter will first look at

D. Slama (✉)
Ferdinand Steinbeis Institute, Berlin, Germany
e-mail: dirk.slama@bosch.com

© The Author(s) 2023
D. Slama et al. (eds.), *The Digital Playbook*,
https://doi.org/10.1007/978-3-030-88221-1_14

co-creation in the context of AIoT before discussing the more traditional sourcing approach.

14.1   Co-Creation

Co-creation between different companies can be an attractive alternative to the more traditional buyer/supplier relationships. This chapter examines different co-creation models, specifically from an AIoT perspective, before bringing in expert opinions from different perspectives, including enterprise, start-ups and venture capital (Fig. 14.1).

14.1.1   Why AIoT & Co-Creation?

What are reasons for co-creation and partnering in an AIoT initiative? On the more strategic level, branding, access to an existing customer base, the exiting global footprint of a partner, access to physical assets or outlets (e.g. repair stations), or domain know-how and existing applications can be good reasons.

Since data are the foundation of any AI-based business model, this can play another key role. Often, data-related co-creation is about the federation of data from two domains, with the expectation that the sum here is larger than the parts. For example, co-creation partners can federate IoT-generated data from multiple machines in a manufacturing setting to support a more holistic OEE perspective for the end-customer.

The development of AI/ML algorithms can also be an interesting area for co-creation. Of course, this will often also be closely tied to the data side of things. For example, the massive costs of developing AI for autonomous driving make some OEMs consider a co-creation/strategic partnership approach in this area.

The need to combine different, highly specialized technologies in the context of AIoT can be another reason for co-creation and partnerships. This can be the combination of different IT enabling technologies, or the combination of IT and OT technologies. Especially with IT/OT, it is often the case that the required expertise is not found in a single company.

Platforms are another interesting area for co-creation. Especially in the area of platforms that need to combine know-how from different industry domains, or provide some kind of data federation, this can make sense. Co-creation here is not limited to technical co-creation. For example, if partners decide to create a platform as a joint venture because they want to pool data, this can also be seen as a form of co-creation (Fig. 14.2).

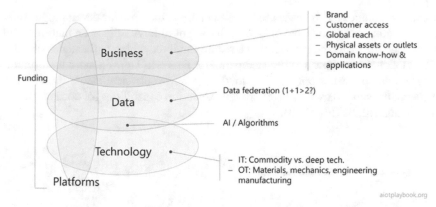

Fig. 14.2 Why AIoT & co-creation?

14.1.2   AIoT Co-Creation Options

Because many people have a different understanding of what co-creation actually means, the following provides a short discussion of common patterns.

In some cases, companies that are actually in a traditional buyer/seller relationship will extend this relationship, e.g., by creating a press release about a "strategic development partnership". Or, they will apply value-based pricing, i.e., the seller is participating in the business success of the buyer. In the case of smaller suppliers, it is also common that the buyer insists on a source-code escrow to have access to the sources in a worst-case scenario or even insists on a stock right of first refusal. These are examples that we would NOT consider to be co-creation, because they are too close to a traditional sourcing relationship.

Co-creation can be more technical or more focused on joint Go-to-Market (GTM). Technical co-creation usually means that – at least – two companies are combining their Intellectual Property (IP) in order to create something new. For example, this could be a company with deep industrial domain know-how partnering with another company that has deep AI experience. Another form of co-creation is focused more on combining two existing offerings into a new offering, but without a deep technical integration. This could mean truly combining two existing offerings or actually selling one company's offering via the channels of the other. Of course both approaches can be combined.

This does not have to be limited to 1:1 partnerships but can also lead to multi-party ecosystems.

Platforms are another area for co-creation. A partner platform will allow partners in a closed ecosystem to work together. For example, a platform for an industrial robot could allow partners of the robot manufacturer to submit applications, which are then operated in a semi-sandboxed environment. Because all the partners are well known and trusted, this approach will be possible without a fully secure execution sandbox. This makes sense, especially if the cost for developing a sully secure sandbox is prohibitive or technically impossible. A fully open platform will have made the investment to develop a secure sandbox, which will enable onboarding of unknown and per se untrusted partners. This will be even more important if apps are deployed not only in the cloud but also on physical assets. In an AIoT scenario, the sandbox will need to provide execution capabilities for code as well as AI/ML algorithms (Fig. 14.3).

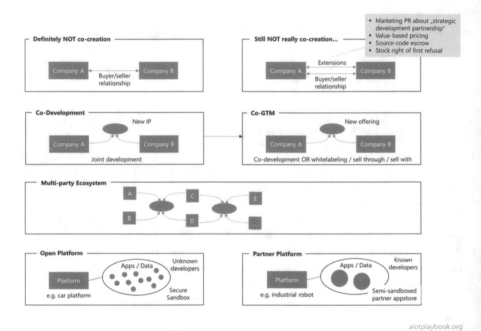

Fig. 14.3 AIoT co-creation options

14.1.3 Expert Opinions

The following interview provides insights on AIoT and co-creation from different perspectives, including large enterprises, venture capital, and research. The experts are:

- Jean-Louis Stasi, Senior Vice President, Strategic Partnerships with Startups, Schneider Electric
- Dennis Boecker, Global IoT Innovation Lead at Bosch
- Ken Forster, Executive Director at Momenta (VC)
- Prof. Heiner Lasi, Director Ferdinand-Steinbeis-Institute

Dirk Slama: *Jean-Louis, how is Schneider Electric managing co-creation?*
Jean-Louis Stasi: *Schneider Electric has twenty lines of business in different areas of energy management and automation – each one roughly with revenues of a billion Euros. Each has its own market environment and its own R&D roadmap. Each is facing different regulatory requirements, which has a huge impact on innovation as well. Take the highly regulated electricity grid market versus fast moving areas such as data centers. Each business has its individual set of conditions in which they are operating, and that is nothing new. However, this defines the appetite for co-creation for each business differently and the different ecosystems they are focusing on. The starting point is always the same: what is the problem that the business is trying to solve? They see the opportunity and the value of the market, but they do not have the required skills and resources internally. They understand that this is going to happen in the next two to three years – so they know they have to move fast. Cyber-security is a good example. That is the next big thing for every industry customer in the world. If I do it myself, it will take me six, seven years. The alternative is to find a start-up as a partner and then scale up together. Of course, going into such a partnership mode is a structural and strategic decision.*
DS: *So the main rationale for co-creation is time-to-market?*
JLS: *No, not necessarily. There are other factors, such as a better understanding of market cycles or specific use cases. Another aspect is the management of business model evolution vs. technological game changes. So I do not want to emphasize time-to-market only, but the point is it has to solve a real-life problem. It has to solve a big problem and has to happen in the next three to five years. Those are the three key conditions.*
DS: *Dennis, what is your take on this from the Bosch perspective?*
Dennis Boecker: *I agree in general with what has been said already. What I would probably emphasize more is that we're coming from our strategic search fields. Many of these strategic search fields are cross-divisional topics, because major trends such as mobility, construction and building technologies cannot be limited to a single line of business. The strategic search fields agree with the executive management, and are aligned with the different lines of business. The interesting question then is how you address the different strategic search fields. We have identified four different ways of doing this: Corporate Product Innovation, Strategic Partnerships, M&A, and Start-up Co-Creation. Corporate Product Innovation is an internal approach in which we focus on building up our own Intellectual Property (IP) and capabilities. Strategic Partnerships are for big projects, taking more a consortial approach. M&A for us is the enhancement of our own IP and our core capabilities. Finally, Start-up Co-Creation focuses on very concrete portfolio elements, or very specific problems we need to solve. Here, time-to-market is usually more important than IP. Say, for example, we are involved in a large bid ourselves, where we have identified a gap in our own portfolio. A start-up is able to address this very quickly. We can even work on several portfolio elements with different start-ups. This is a good way of creating an ecosystem. You share a problem, and then you start working on concrete proj-*

ects, either in a more strategic partnership, with start-ups, or even a combination thereof. Co-creation for me is not a one-to-one model, where two companies work together to create a specific piece of IP. We truly want to build open ecosystems around our strategic search fields, including multiple partners. And then applying any of the four collaboration models which I just explained, depending on the situation.

JLS: *Interesting. Schneider Electric has a very decentralized model of control. Each line of business is pretty much autonomous in their selection of partnership models and specific partners. So there is no centralization of this element. At the end it produces a central result, where each business has its own iteration on that. An important element is the intuition between a start-up and the larger structure. A start-up is not necessarily three guys in a garage. It can already be scaled by a team of 200 engineers, which have already scaled to a certain level of product-market fit. As we grow more confident in the potential for scaling their business, they become an M&A target for us. However, this is a continuous process, and it is not decided in the early stages. It takes some time for a partnership to evolve into something more, before it becomes strategically impactful for us at large and we are prepared to make a move in the direction of M&A. Another important aspect in this is the geographical footprint. Because what we see is of course markets are evolving at a different speed, depending on where the innovation is happening. Is it US? Is it Europe? Is it China? And each market also has its own way to do things. So we also have this dimension that we need to integrate into our model because as you develop that globally, there is no one way to do that things. Each kind of region has its own specifics, depending on the maturity and culture.*

DB: *I agree with you on the necessary evolution of partnerships. There is definitely a chance that the portfolio approach and the respective relationships with the ecosystem develop from one quadrant into the other. But in the very beginning, I think you need to be very clear if this is something to share with the ecosystem or if it is rather something that you want to own the IP ultimately yourself. That is something that you need to be very clear about.*

DS: *Let us switch the perspective from the enterprise side to the start-up side. Ken, Momenta is an investment fund that specifically focuses on AIoT companies. What are you telling the companies in your portfolio: look for a buyer-seller relationship with large enterprises, or focus on strategic partnerships?*

Ken Forster: *We see it as an ecosystem of many-to-many, with many actors contributing to the overall value of solutions. Let me first outline the actors and then I can reflect on the role Venture Capital plays. I will generalize four key actors in the AIoT ecosystem: incubators, innovators, incumbents, and the implementers. As investors, we broadly operate as an incubator working to grow young companies that are often innovators. The large strategic Operational Technology (OT) players are generally incumbents. Finally, we have the end users, implementers, those leveraging our collective technologies and tools to create business outcomes. Between these actors, Venture Capital sit at the intersection of innovators, incumbents and implementers accelerating the velocity of innovation. Early* stage companies often do not have the resources or experience to dance with the elephants of industry, so we bridge the gap, investing behind the innovators to accelerate their capabilities to play in the larger ecosystem. While our initial value is in providing seed to later-stage capital, the larger value is often the acceleration we bring via our AIoT ecosystem networks. As an example, we organized a consortium of four of our portfolio companies last year to demonstrate an intermodal container location and condition monitoring demonstration for the U.S. Department of Defense. The pilot demonstrated the use of AIoT technologies to track the location of containers through the full journey of shipment while recording the condition of the cargo within those containers, including security. This was truly a team effort with each portfolio company providing a critical piece of the solution. The DOD in this case was the implementer actor, our portfolio companies the innovators with Momenta acting as the incubator. Of course, once product-market fit is demonstrated, the incumbents will often play the strongest role: providing the 'voice of the customer' backed by their own size and momentum to scale up these innovations to the enterprise scale. In summary, we operate at the Venn of innovation – bridging the innovation of startups with the enterprise scale of large industrials. Our tools are capital, securing key leadership and teams, and activating partnerships early and often with companies such as Bosch, Schneider Electric and other market leaders.

DS: *Thanks. Now let us look beyond AIoT-enabling technologies for a moment. Heiner, another key ingredient to AIoT is data. In your research, you are focusing on ecosystems for data sharing to support different AIoT use cases. Care to explain?*

Heiner Lasi: *As you said, an AIoT use case needs data to operate on. If these data are created and processed within a single company, everything is fine. However, in many cases, you will need to cross these boundaries. Either because you need to combine data from different IoT-objects/companies, or because companies need to access data from other IoT-objects/companies in order to create business value. Let us take a logistics company that is running a fleet of forklifts. The initial IoT business cases here were very much about operational efficiency. This was all internal. But now we are seeing "… as a service" business cases. This means that logistics companies and forklift manufacturers must share operational data. In addition, insurance and finance companies are becoming involved in these business cases. On the technical level, emerging concepts such as cooperative Data Spaces can help here. However, the issue here is not technical integration but rather the creation of a partner ecosystem that is developing the required levels of trust to share operational data between the partners. This is not about technical trust; this is about trust on the business and even the human level. Transparency is important. This is what we are trying to address with our concept of data coops. These data coops are a manifestation of a data-centric ecosystem, with clearly defined rules of engagement, and a data space as the underlying technical platform. The rules of engagement must address how – for example – the logistics companies, the forklift manufacturers, and the financial companies are providing and accessing data, and how the benefits are shared.*

While AIoT is an important technical enabler in all of this, it really comes down to creating an ecosystem of partners who trust each other. And it takes time to develop the required level of trust.

JLS: *This is an extremely critical point. What we are seeing at Schneider Electric is that there is a very strong relationship between the time you connect an asset with a sensor, and the time you get tangible business value out of it. Time-to-value is of the essence. I think we – all the industrial companies – have a long list of IoT use cases which have created zero value. Anything beyond three months, you lose the momentum, you lose the sponsorship. If you are thinking too broad, try to integrate too many stakeholders, things get too complicated. Every factory you talk to is adding new requirement. The conditions in the factories are different. The cloud infrastructure is different. The level of knowledge is different. It never ends. And suddenly you are caught in the "pilot dead end".*

To Heiner's point about the data: the challenge is to ensure that these data are turned into a tangible benefit for the customer in a very short amount of time. To achieve this, you have to maintain OT leadership in the team. You must give them the value, which means you also must give them the leadership. If you give the leadership to the IT guys, it is finished. That is our experience.

HL: *I agree. However, this is not only about IT vs OT. This is also about creating a link between business experts from different industry domains, such as OT and financial services, e.g., to enable "as a service" business models. In this example, you need to combine the experience of running the OT side with the experience of understanding the inherent financial risks.*

JLS: *Let us take supply chain visibility, e.g., in the food and beverage industry. This is a good example where we need to unify data to obtain increased visibility. It took them years to realize that yes, it makes sense to share data, because we have the same suppliers and we want to make sure that this food product truly is coming from the right place. It took perhaps five years to reach this level of trust in this ecosystem. But then we are coming back to time-to-value: People do not have five years to get to the outcome. They have to create value in the next six months. So there is this big tension in the AIoT space regarding the time it takes to build trust. Maybe eventually there will be a technology such as blockchain to intermediate the trust, but I think thus far this has not yet scaled beyond one or two verticals. I want to reiterate this, because I think AIoT has been failing in many, many cases because of this lack of value in a given time frame. In addition, I think we are still on this journey where it takes more value creation to justify a vertical integration, end-to-end.*

DS: *Does this only apply to what we have been calling the long tail of AIoT, or are we also talking about the short tail of AIoT here?*

DB: *Even for the high-impact products on the AIoT short tail, you need to ensure that you are delivering value creation along the way. If you have this huge AIoT short tail opportunity which you know will take you three to five years to deliver, you better make sure that every couple of months on the way to it you show tangible value creation. This is what I was referring to earlier on regarding portfolio achievements and strategic partnerships. On this level, it will take a longer time, and you cannot expect to get the full outcome in three months, because it's just too much and too complex. However, you need those complementary, smaller items along the way to show that AIoT makes sense. Admittedly, with physical products involved, finding a stable Minimum Viable Product is much more difficult, but still...*

KF: *In the past, AIoT products often required full stack solution development. As an example, Nest had to develop the full stack of software, hardware and connectivity to build a smart thermostat. Today, this development is more horizontal allowing solution developers to choose from best of breed components leveraging standards. In this momentum from highly vertical development toward more horizontal solutions, there is even more opportunity to work with an ecosystem of companies that are aligned around domain specific use cases.*

HL: *We see a similar trend for data coops. Usually, they are initially clustered around domain-specific use cases. Therefore, the data coop with the underlying data space is the platform, with the different use cases representing long tail opportunities. Because there is an upfront investment in setting up the initial platform, it is key to immediately show value creation with the first use cases.*

JLS: *If you are talking about high-impact products such as the first iPhone, I agree. They take significantly more time and higher investments. Even if we are talking about IT/OT integration, it is important to understand if you are talking about the device manufacturer, or the implementer. If you are talking about the manufacturer of a new AIoT-enabled device, I agree with your characterization. However, if you are talking about the implementation side, where you are addressing many factories in different locations, you need to do this in a very consistent and predictable way. You will need a very simple and easy way, so that every operational team can implement this in less than three months. Than you can reach massive scale because you are making things so simple. Then, it becomes like the SaaS (Software-as-a-Service) model. This is where AIoT today starts to become relevant for industrial companies, because end users are able to scale this up themselves.*

DS: *Thank you all!*

14.1.4   Tradeoffs

To conclude the discussion on co-creation and strategic partnerships, let us take a short look at the pros and cons. For example, some of the risks and drawbacks include:

- Technical, legal and commercial complexity, as well as generally increased stakeholder complexity leads to a significant increase in project risks
- Colliding interests can lead to failure or at least unbalanced partnerships
- Limited visibility and lack of transparency can impose significant risks

Some of the benefits include:

- Significant business and technical synergies
- Unlocking creative potentials that would not be possible within a single company
- Access to markets otherwise not easily accessible
- Combining speed and agility of startups with global reach and execution capabilities of incumbents

Before making a decision on co-creation vs. traditional sourcing, these factors need to be carefully weighed.

14.2 Sourcing

The acquisition of the required technologies and resources is probably one of the most critical parts of most AIoT projects. Many project leaders – and many procurement departments – do not have much experience in this space, which is why this part of the book aims to provide a structured approach to the problem (in combination with a set of useful templates) (Fig. 14.4).

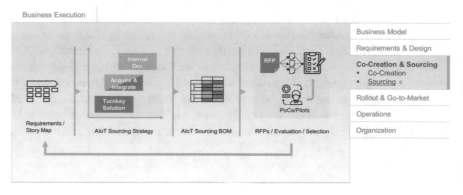

Fig. 14.4 Sourcing

The approach described here covers typical sourcing challenges, introduces a generalized sourcing process for AIoT products/solutions, discusses *make vs. buy vs. partner*, introduces the concept of an *AIoT Sourcing BOM*, helps define vendor selection criteria, covers RFP document creation and management, and finally looks at vendor selection.

14.2.1 Challenges

Before looking at the details of the sourcing strategy and process, we must first understand the challenges associated with AIoT sourcing and procurement. An AIoT project is often a complex undertaking. On the business side, many different stakeholders must be aligned, contradicting requirements must often be managed, and existing business processes will have to be re-engineered. On the technology side, a multitude of new technologies and methodologies must be made to work together. In the case of a line-fit solution, existing or new manufacturing capabilities must be aligned with the needs of the AIoT solution. Finally, the solution roll-out and service operations must be prepared and managed.

So what can go wrong? A lot, especially if project management does not pay close attention to the digital supply chain. Selecting the wrong vendor or the wrong technology for all or parts of an AIoT solution can have ripple effects that put the entire project in danger. The same applies to over-specifying or underspecifying what is needed. Poor implementation services or badly defined SLAs (Service Level Agreements) can lead to bad user experience and stability problems. If these problems are only determined after the roll-out, this can put the entire business at danger. The list of sourcing-related challenges also includes issues with adapting to change, allowing poor quality for lower costs, ignoring the costs of time, ill-defined sourcing and procurement processes with unclear responsibilities, project management issues, complex organizational dependencies, and loopholes in contracts.

Especially for industrial companies, dealing with AIoT-related topics from a sourcing point of view can be challenging:

- How can agile development be supported with a matching pricing model and contracts?
- How can we deal with new paradigms such as AI and the required SLAs?
- Can AI-based solutions be treated like software-based solutions from a sourcing point of view, or do they need a different approach?
- How can dependencies between different suppliers be managed, e.g., for hardware and software?
- How can vendor lock-in be avoided due to 'accidental' technical dependencies?

RFPs (Request for Proposals) play an important role in many sourcing projects. Depending on the chosen sourcing strategy, a number of different RPFs will be required, especially if different technologies and resources will be acquired from different suppliers. One should not underestimate how unpredictable and difficult to manage RFP projects can be and how often they miss their deadlines. Carefully aligning the required RFP projects with your development plans will be a key success criterion for your project. One aspect here simply is the timelines for running the RFP and securing suitable vendors. Especially in larger enterprises, another aspect is the complexities of sourcing decisions in complex political environments.

14.2.2   AIoT Sourcing Process

The first important step towards successful technology and resource acquisition is to define a high-level process, which needs to be aligned with all key stakeholders: AIoT project team, procurement, legal, and often senior management. The process proposed here is based on the assumption that it will be centered around a Request for Proposal (RFP), and has five main elements: sourcing strategy and planning, RFI (optional), RFP creation, RFP distribution, and AIoT vendor selection (Fig. 14.5).

Fig. 14.5 AIoT sourcing process

Procurement strategy and planning need to look at the most important aspects of the AIoT solution (including stakeholders, scope, and requirements), as well as the implementation project (timeline, key milestones, and budget). As part of the sourcing strategy, the make vs. buy question must be addressed. Depending on the outcome of this decision, the creation of a specialized *Sourcing BOM* (a breakdown of all required elements of the solution) should be considered. Furthermore, vendor profiles, sourcing criteria, and the actual sourcing process (including timelines) should be defined.

During the RFP creation phase, a concise RFP document must be created, reviewed with all internal stakeholders, and often formally approved.

After completion of the RFP document, it will be distributed to the target vendors. In some cases, it will also be made publicly available. Managing the RFP process will usually involve a structured Q&A process with all interested suppliers.

Finally, the vendor responses must be evaluated. Often, vendors are invited for individual vendor presentations. Based on this information, a first set of vendors can be preselected. In some cases, smaller Proof-of-Concept projects are done with these vendors. Based on the outcomes of the PoCs, a short-list can be created. Often, the last few vendors are then asked to do a more extensive pilot project. Based on the technical and functional evaluation, as well as extensive price negotiations, the final selection is then done.

14.2.3   AIoT Sourcing Strategy

Defining the sourcing strategy is an important first step. This section will cover strategic sourcing options (make vs. buy. vs partner), the AIoT Bill of Materials, the AIoT Sourcing BOM, and finally the alignment with the development schedule.

14.2.3.1   Strategic Options: Make vs. Buy vs. Partner

The decision for a specific sourcing strategy is fundamental and will shape your AIoT-enabled business for the years to come. Giving up too much control over the production process for a strategic product can be as problematic as investing too many own resources in the development of commodity technologies and failing to build truly differentiating features on top.

So what are the options? For the purpose of our discussion, we have identified three strategic sourcing options:

- Internal Development: This option basically assumes that only commodity technology such as middleware or standard hardware components will be externally sourced, but all custom development (including software and hardware) will be done internally.
- Acquire & Integrate: This option assumes that only the high-level design and component integration will be done internally, while all subcomponents (hardware, software) will be acquired from external sources.
- Turnkey Solution: This option assumes that an external provider will be selected to provide a complete solution or product, based on the requirements defined by the ordering organization. This can either be a complete custom development, or the customization of a standard solution/Commercial-Off-the-Shelf product. Typically, in this case, the supplier is responsible not only for the implementation, but also for the design.

These three options are only examples. Other options, such as co-innovation or Build-Operate-Transfer, can also be interesting. However, these three examples should provide a good starting point for the discussion in the following (Fig. 14.6).

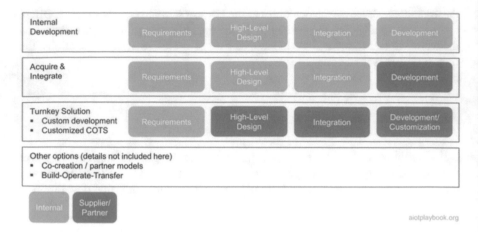

Fig. 14.6 AIoT sourcing options

So how to decide for the right sourcing option? One key factor is the strategic relevance of the AIoT-based product or solution. An auxiliary system with little direct impact on the business could probably best be acquired as a turnkey solution. A strategic product that will be responsible for a large part of future revenue will most likely require much more control over the product's design and value chain, and thus lend itself to the custom development option. The same could hold true for an AIoT solution that controls parts of an enterprise's core processes.

Other factors that must be taken into consideration include the following:

- Organizational capabilities: Does your organization have a proven track record in hardware and/or software development? And how about AI and Data Science?
- Resource availability: Do you have the required resources available for the required time period? And is it the best use of these resources?
- Could you build it fast enough?
- Could you build it good enough?
- Need for control: How much control does your organization need over the design and value chain?
- Would building it internally allow cost reduction (e.g., by utilizing own manufacturing lines)?
- Do you want to keep building/maintaining it yourself after the launch/SOP?
- How mature is the supplier market?
- Is there an opportunity for a strategic partnership here?
- Is a well-known supplier brand a potential differentiator?

In many cases, the Make/Buy/Partner question cannot be answered for the entire product or solution but needs to be broken down to different components (see discussion on the Sourcing BOM below). To answer the Make/Buy/Partner question for a complex AIoT solution, it is often important to first understand the complete breakdown of the solution. This is examined in the discussion of the *AIoT Sourcing BOM* below (Fig. 14.7).

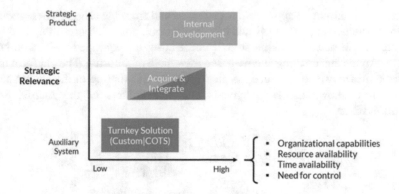

Fig. 14.7 Sourcing strategy decision

14.2.3.2   The AIoT Bill of Materials

In manufacturing, the bill of materials (BOM) is used for planning the purchasing of materials, cost estimation, and inventory management. A BOM is a list of every item required to build a product, including raw materials, subassemblies, intermediate assemblies, subcomponents, and parts. It usually also includes information about the required quantities of every item.

There are usually different, specialized BOM-types, including:

- Engineering BOM: developed during the product design phase, often based on Computer-Aided Design (CAD) data. Lists the parts and assemblies in the product as designed by the engineering team
- Manufacturing BOM: Includes all the parts and assemblies required to build the finished product. Used as input for the business systems involved in ordering parts and building the product, e.g. ERP (Enterprise Resource Planning), MRP (Materials Resource Planning), MES (Manufacturing Execution System)
- Sales BOM: used during the sales phase, provides details of a finished product prior to its assembly

Given the potential complexity of an AIoT project, we propose the creation of a *Sourcing BOM* as the foundation for the sourcing process. In the following, we start with a discussion of a generic AIoT BOM, followed by the introduction of the Sourcing BOM.

14.2.3.3   Example: ACME Smart Shuttle

To provide a meaningful discussion of the BOM concept for an AIoT product, we use the ACME Smart Shuttle example. ACME Smart Shuttle Inc. is a fictitious company offering a platform to manage shuttle services for schools. Instead of using a fixed bus network and fixed bus schedule, ACME Shuttle utilize AIoT to offer much more on-demand service to students. Instead of using fixed bus stops, virtual bus stops are introduced that can change during the day, depending on demand. Students are using a smartphone app to request a ride to and from the school. These requests are then matched against the virtual bus stop system, potentially resulting in the ad-hoc creation of new, virtual bus stops. Shuttle buses are equipped with an on-board unit to provide bus tracking and AI-based in-vehicle monitoring. The platform in the backend utilizes AI to optimize the pick-up order and routing of the shuttle buses. Figure 14.8 shows the key elements and stakeholders of the ACME Smart Shuttle system.

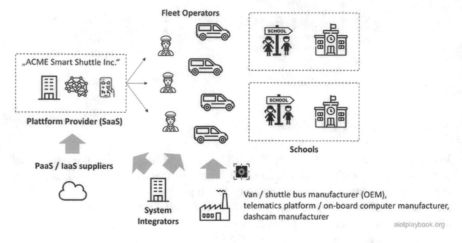

Fig. 14.8 Example: supply chain of our shuttle bus system

To return to the BOM discussion, the starting point for the creation of a basic BOM structure for an AIoT product or solution is usually an analysis of the architecture design. Figure 14.9 shows an example of the high-level architecture design of the ACME Smart Shuttle solution. Additionally, listed are examples of resources required for implementing key elements of the system.

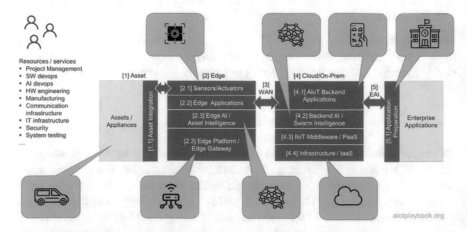

Fig. 14.9 Solution architecture as the starting point for BOM breakdown

14.2.3.4   Creating the AIoT BOM

A BOM is typically a hierarchical structure; in our case, the 3–5 high-level areas of the solution architecture should form the first hierarchy level of the BOM. Note that this BOM will include not only hardware, but also software elements, as well as network infrastructure. In reality, the BOM for such a project might be comprised of multiple, specialized BOMs. The example below indicates how a high-level architecture design – such as the one for the ACME Smart Shuttle example from before – can be mapped to the initial BOM.

Thinking about required resources in terms of a BOM will be unusual for people from the software world. However, the benefit of including not only hardware and physical elements in the BOM structure but also software and virtual elements is that the BOM provides a holistic view of the entire system. This can be used not only for the Sourcing BOM but also from the point of view of dependency management, tracing of BOM elements from a security point of view, etc. (Fig. 14.10).

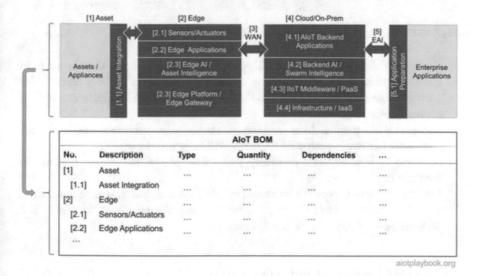

aiotplaybook.org

Fig. 14.10   AIoT sourcing BOM: creation

14.2.3.5   Make vs. Buy Breakdown

For most AIoT systems, the make vs. buy (vs. partner) decision cannot be applied to the entire system. Instead, it must be applied to different entries in the AIoT BOM. Fig. 14.11 shows four different scenarios:

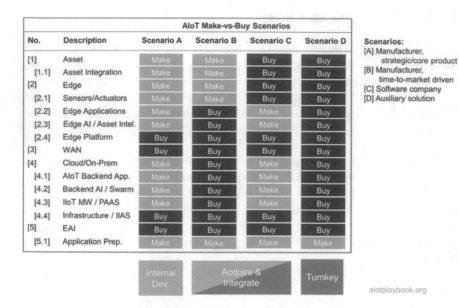

Fig. 14.11 Sourcing BOM with different sourcing scenarios

- Scenario A is a manufacturer working on a strategic new core AIoT product. In this case, most BOM items will be custom made internally. Only some items such as Edge Platform, WAN, cloud infrastructure and EAI middleware, will be sourced externally.
- Scenario B is a manufacturer working on a time-to-market driven project. In this case, only hardware-centric BOM items will be sourced internally.
- Scenario C is a software company that takes nearly the inverse position to scenario B.
- Finally, scenario D assumes an auxiliary AIoT system, which will be sourced as a turnkey solution. Only the preparation of existing applications for integration with the new system will be done internally.

14.2.3.6   ACME Smart Shuttle: Outsourcing AI?

ACME Smart Shuttle, Inc. sees AI as a key enabler to build highly differentiated product features with a strong customer appeal. Consequently, the team has performed an assessment of the best uses of AI in the system design. The most promising AI use cases have been discussed with the procurement team as part of the BOM creation. A summary of the make vs. buy vs. partner decisions that have been made is summarized in Fig. 14.12.

BOM ID	AI-enabled system components	Description	Sourcing Decision	Reason
4.2.1	Shuttle routing	Dynamic routing of shuttle buses, based on user demand; includes management and optimization of virtual bus stops	Build-operate-transfer	Key part of the system, main differentiator – but no own experience in hiring and managing a team with the required skills; also, required experts not easily available on the market
4.2.2	Shuttle ETA forecast	Estimated time of arrival (ETA) for shuttle buses at differend virtual bus stops (to be matched against shuttle routing)		
4.2.3	Driver shift planning	Long, medium and short term planning of shuttle bus driver's working shifts, depending on demand forecast (student timetables, holidays)		
2.3.1	In-vehicle surveillance	Video data analysis to track students entering/existing the shuttle busses, automatic detection of violence and vandalism	Buy / co-creation	Not a main differentiator, but some functions potentially not readily available (e.g. vandalism detection)
2.3.2	Vehicle maintenance	Use of telematics data for predictive and prescriptive vehicle maintenance	Buy	Not a differentiating feature

Fig. 14.12 Outsourcing AI?

Three AI-enabled components have been identified as particularly important to the system: Shuttle routing, shuttle ETA forecasting and driver shift planning. Ideally, these three components should be developed in-house to retain full control and ensure constant optimization. However, the analysis has shown that the engineering management team has no experience hiring and managing a team with the required AI skills; furthermore, the required AI experts are not easily available in the market. Consequently, the decision was made to opt for a build-operate-transfer model: the development and operations support for these components will initially be outsourced. Medium- to long-term, ACME Smart Shuttle will then take over the team from the external supplier to become part of the in-house organization.

For AI-enabled in-vehicle surveillance and vehicle maintenance, the decision was made to buy these components because they are not strong product differentiators and commodity solutions should be available with potentially one exception: the automatic detection of violence between students or even vandalism. For this particular feature, a co-creation model could be envisioned, assuming that there would be a market for such a feature beyond the business scope of ACME Smart Shuttle.

14.2.3.7   AIoT Sourcing BOM

The next step is to turn the generic AIoT BOM into an *AIoT Sourcing BOM*. The first thing that needs to be looked at in more detail are the required quantities:

- For hardware components deployed on the assets, the required quantities will depend on the number of assets to be supported. This again will depend on the business plan. This means most likely the correct strategy here will have to foresee different options, like a minimum and a maximum amount required.

This will have to be mapped to different contractual options with the suppliers.

- Additionally, for software licenses, the number of clients often plays an important role. In the case of AIoT, clients can either be human users or assets. Again, this will depend on the business plan and require some flexibility to be built into the sourcing contracts.
- Finally, for custom developed software, the Sourcing BOM will sometimes have to include an estimation of the required development resources (number of developers, availability). Alternatively, this is an estimation that can come from suppliers, based on the requirements.

Next, for each item in the Sourcing BOM, a sourcing decision will have to be made. Sourcing options typically include internal development, management consultancies (e.g., for project management), System Integrators, Commercial Off-the-Shelf Software Vendors, Cloud infrastructure providers, engineering companies, manufacturers, and telecommunication companies.

A key decision for each element in the Sourcing BOM is the make vs. buy (or partner) decision. This decision will depend on a number of different factors:

- Strategic importance of AIoT Solution as a whole and the contribution of each BOM item individually
- Internal capabilities: is this something your company can realistically do itself?
- Availability of internal resources
- Timing: who can deliver within the required time frame?
- Brand considerations: will having a certain brand for a specific subcomponent improve the overall value of the product?
- Overall partner strategy: does it make sense to utilize some companies not only as suppliers, but also as potential additional sales channels?

Once quantity and sourcing strategy information has been added to the Sourcing BOM, the schedule perspective needs to be added as well. This needs to be carefully aligned with the development schedule to avoid roadblocks on the development side.

Finally, it is important to note that in a complex AIoT project, not all required solution elements may be known from the beginning (or they might be subject to change). Agile development methodologies are designed to address volatile requirements and solution designs. However, from a sourcing point of view, this is obviously very problematic. Frequent changes to the Sourcing BOM will result in loss of time and potentially even spending money on the wrong things (Fig. 14.13).

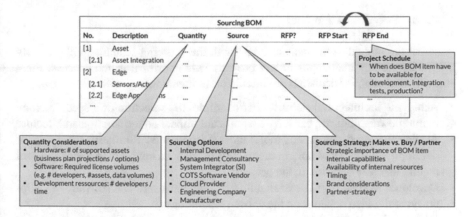

Fig. 14.13 AIoT sourcing BOM: refinement

The following provides some examples of typical elements of an AIoT Sourcing BOM specifically from the point of view of AI- and IoT-related components.

AI-specific Sourcing BOM Elements

The following are some examples of typical, AI-specific elements of an AIoT BOM:

- AI platform, including AI-specific hardware and middleware – for use in the cloud/on-premises backend, or the EDGE layer
- Functional components requiring resources with AI-specific skills, including the AI engineer, data scientist and AI DevOps engineer
- Outsourced data labeling services, e.g., for manual image classification; beware that transferring images with personalized data to other countries for such processing services can be prohibited by local regulations.
- AI-specific QA, testing and validation services

IoT-specific Sourcing BOM Elements

The following are some examples of typical, IoT-specific BOM elements:

- IoT-related cloud infrastructure
- EDGE infrastructure (hardware, software)
- Resources with IoT-specific skills, e.g., embedded hardware or software development, AIoT project management, etc.
- Telecommunications services, e.g., a global IoT network from a telco carrier or an MVNO
- Security-related infrastructure, testing services, operations services and skilled resources
- Operations services and support

14.2.3.8 Schedule Alignment

Aligning the agile development schedule with the sourcing schedule will probably be one of the key challenges in any project. This is critical to the success. Final sourcing and supplier decisions are often a prerequisite for:

- Achieving architectural stability: For example, the selection of a specific cloud or middleware platform can have a significant impact on the solution architecture
- Availability of development tools and environments: Similarly, the setup of development tools and environments will usually be supplier-specific, and will require an early decision in the project
- Developer availability: The availability of both hardware and software developers typically also depends on the chosen technology
- Infrastructure setup: Additional infrastructure such as an AI environment or a security framework will again depend on the final sourcing decision
- Hardware development: Finally, any hardware-specific development will also require sourcing decisions, e.g., for processors, boards, or communication modules

Figure 14.14 highlights the potential dependencies between the agile development schedule and the sourcing schedule.

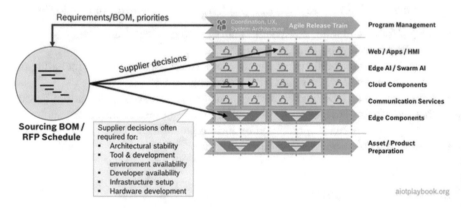

Fig. 14.14 Schedule alignment

14.2.4 General Considerations

Before starting the RFP process, a number of other general considerations must be made, including the required SLAs and Warranties, pricing models, and vendor selection criteria.

14.2.4.1   SLAs and Warranties

A critical decision in the procurement process is the type of contract that is aimed for, especially for any kind of custom development:

- Service contract: Typically, time and material
- Contract work: Typically, includes SLAs, maintenance commitments, warranties, etc.

In many situations, the latter will be particularly important for an AIoT solution. Warranties typically ensure that a service will perform in accordance with its functional, technical and business specifications. Service Level Agreements (SLAs) offer performance metrics and details on the specific consequences of a provider who is failing to meet those standards.

Typical SLAs in IT projects include:

- Service availability: Specifies the amount of time a service is available, e.g. 99.99% (which would imply ~88 hours of average annual downtime)
- Defect rates: Quantification of allowed error rates in a service
- Defect resolution: Addresses the speed by which problems are addressed
- Security: Addresses the security of the service
- Business results: Address the business perspective, e.g., as business process metrics

Figure 14.15 shows some examples where this is applied to an AIoT Solution.

AIoT BOM Item	Example SLAs
AI (Asset or Swarm Intelligence)	Compliance with intended business functionality, e.g. matching accuracy in %
Edge Platform	Event processing, e.g. #events / second
IoT Business Logic (Edge or Cloud)	Compliance with design specifications
IoT LAN	Coverage of required regions, network latency, throughput

aiotplaybook.org

Fig. 14.15   Sourcing BOM SLAs

14.2.4.2   ACME Smart Shuttle: SLAs for AI?

The ACME Smart Shuttle had previously identified three key components for the system, which utilize AI. The decision was made to apply a build-operate-transfer model as the sourcing strategy for these three components. This means that component development will initially be sourced externally, with the goal to then in-source the team over time. To ensure that the external team meets the requirements, a set of SLAs have been defined. These SLAs differentiate between functional and non-functional aspects. The functional SLAs are specific to the individual components, while the nonfunctional SLAs in this case apply to all three components. Figure 14.16 provides an overview.

BOM ID	AI-enabled system components	Description	Functional SLAs	Non-functional SLAs
4.2.1	Shuttle routing	Dynamic routing of shuttle buses, based on user demand; includes management and optimization of virtual bus stops	• Shuttle bus utilization (avg. number of passengers per driven mile) • Average waiting times for students • Handling of peak demand times • Average idle time of shuttle buses	• Monthly cloud compute costs • Cloud compute costs per mode re-training • Offline training duration • Online response times
4.2.2	Shuttle ETA forecast	Estimated time of arrival (ETA) for shuttle buses at differend virtual bus stops (to be matched against shuttle routing)	• Accuracy of ETA forecasts	
4.2.3	Driver shift planning	Long, medium and short term planning of shuttle bus driver's working shifts, depending on demand forecast (student timetables, holidays)	• Matching of shift schedules with actual demand • Worker satisfaction	

aiotplaybook.org

Fig. 14.16 ACME smart shuttle SLAs for AI components

A key issue with SLAs for AI-based components is that AI models usually decay over time, due to changes in the input data. Take, for example, the ETA prediction function for shuttle buses: this function will heavily depend on map and traffic data. If the actual layout of the street grid is changing (e.g., due to construction sites), this will probably require the ETA models to be retrained with the updated map data. This will have to be reflected in the contract as well: The SLA definitions can only apply to models that are regularly retrained.

14.2.4.3 Pricing Models

Another important factor in the sourcing process is the pricing model. In many situations, the customer will define the required pricing model as part of the RFP. However, in some cases, the pricing model can also be defined by the supplier.

In IT development projects, the most common pricing models are Fixed Price and Time and Material. A key prerequisite for a Fixed Price project is a stable, complete and sufficiently detailed requirements specification and Service Level Agreements. If this cannot be provided, then Time and Material might be the only real alternative. Variations of the Time and Material approach are the Dedicated Team approach, as well as Agile Pricing. In Agile Pricing, often a base price is agreed upon, combined with incentives related to the achievement of individual sprint goals. Another pricing option is a model where the supplier participates in the business success of the customer, e.g., revenue sharing ('Outcome-based pricing'). However, getting both sides to agree to a fair sharing of risks and rewards can be a difficult undertaking.

Other elements of the AIoT Sourcing BOM will again require completely different pricing models. For example, the pricing for telecommunications services will often depend on data volumes and other factors. The pricing for custom hardware is likely to depend on individual component prices, as well as volume commitments.

14.2.4.4 AIoT Vendor Selection Criteria

Once it is decided which items from the AIoT Sourcing BOM should be externally acquired, it is important to create a set of clearly defined selection criteria. The Digital Playbook proposes a spreadsheet that includes the AIoT solution in general, nonfunctional requirements, functional requirements, and finally the operations and maintenance requirements. Each of these criteria should be individually weighted, so that later an overall score can be derived for each offer.

In this context, a number of key questions will have to be answered, including the following:

- How important is cost relative to the other areas?
- How important is the ratio between functional and non-functional requirements?
- How important is the vendor evaluation, including strategic fit, financial stability, long-term maintenance capabilities, etc.?

Figure 14.17 shows an example of a spreadsheet containing key selection criteria.

Criteria		Weighting	Min	Vendor: ACME AIoT Solutions	
				Evaluation (1-10)	Weighted
AIoT Solution General		35%			2,45
	Price	15%	7	7	1,05
	Vendor / strategic fit	10%	5	6	0,6
	Completeness of offering	10%	8	8	0,8
Non-Functional Requirements		25%			1,1
	NFR 1.1	5%	6	7	0,35
	NFR 1.2	10%	5	2	0,2
	NFR 2.1	5%	6	6	0,3
	NFR 2.2	5%	5	5	0,25
Functional Requirements		25%			1,5
	FR 1.1	5%	7	8	0,4
	FR 1.2	10%	6	6	0,6
	FR 2.1	5%	7	7	0,35
	FR 2.2	5%	6	3	0,15
AIoT Solution Ops & Maintenance		15%			1,1
	OM 1.1	10%	8	8	0,8
	OM 1.2	5%	7	6	0,3
		100%		Score (1-10):	6

Fig. 14.17 AIoT sourcing criteria

14.2.5   RFP Management

Finally, once the internal alignment is completed, the RFP process starts. This includes RFP document creation, RFP document distribution and Q&A process, and eventually AIoT vendor selection.

14.2.5.1   RFP Document Creation

The creation of the actual RFP document(s) is a critical part of the sourcing process. It is key that an RFP document is as concise as possible, with sufficient detail for any contractual agreement based on it. Any gap or inconsistency in the RFP can be used further down the path by a supplier for re-negotiation or costly change requests. Consequently, the RFP should be written specifically for the situation at hand and not a repurposed, generic document. Typical elements in an RFP include:

- Company name, project name, proposal due date
- Project overview
- Scope of work

  - Functional requirements
  - Non-functional requirements

- Quality criteria
- Submission requirements and process

In many cases, it can also make sense to be transparent about the following:

- Evaluation metrics and criteria
- Budget

For the Scope of Work part, it makes sense to reuse many of the Solution Architecture design artifacts, e.g. the solution sketch, data domain model, component design, etc. However, two key questions must be looked at here:

How many details from the business plan to reveal in the RFP? It can be advantageous to share some details of the business plan with potential suppliers to allow them to get a better understanding of the business potential and thus to make better offers. However, many companies would feel reluctant to share too many details in a document shared with many external stakeholders.

How detailed should the solution design be? Providing a solution design to potential suppliers can be a good way to ensure consistent offers from different contenders, which closely match the requirements. However, it can also be limiting in terms of obtaining different solution proposals with different strengths and weaknesses (Fig. 14.18).

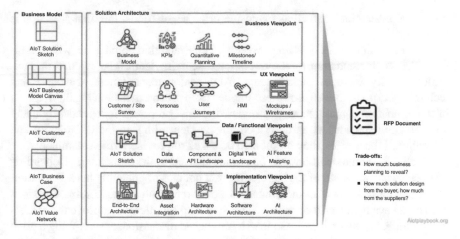

Fig. 14.18 RFP document creation

The Industrial Internet Consortium (IIC) has developed an online tool for creating an RFP for Industrial Internet solutions. While currently lacking the AI perspective, this can still be an interesting tool for anybody creating an AIoT RFP document, at least for the IoT parts.

14.2.5.2   RFP Document Distribution and Q&A Process

After completion as well as internal review and approval, the RFP document is distributed to relevant supplier candidates. In some cases, the RFP might also be publicly made available, if this is an internal requirement.

If the process permits this, the receivers of the RFP are likely to come back with questions. First, almost any RFP will leave some room for interpretation. Second, most suppliers are likely to seek close, personal contact with the acquiring company and the sourcing team. It is important that to run a transparent and fair selection process, the questions from all potential suppliers are collected, and the answers are shared as an update to the RFP with all contestants. This will also help increase the quality and comparability of the offers.

14.2.5.3   AIoT Vendor Selection

As part of the selection process, vendors are invited for individual vendor presentations. Based on this information, a first set of vendors can be preselected. Reference calls can provide valuable insights from other customers of the different vendors. In some cases, smaller Proof-of-Concept projects are done with these vendors. Based on the outcomes of the PoCs, a short list can be created. Often, the last 2–3 vendors are then asked to do a more extensive pilot project. Based on the technical and functional evaluation, as well as extensive price negotiations, the final selection is then done.

The selection process is often overseen by an evaluation committee, which evaluates the recommendation by the operational sourcing team. The evaluation committee usually includes stakeholders from senior management, business and technology experts, as well as representatives from procurement and legal. Members of the evaluation committee ideally review the final proposals independently using an evaluation spreadsheet as described above. Depending on the complexity and criticality of the project, they might also be asked to provide written statements.

Finally, the results will have to be communicated to the contenders. Depending on the internal processes of the buyer, different policies might apply here. For example, it can make sense to communicate not only the result but also some decisions such as the evaluation criteria matrix. This will help suppliers to improve their offers in the future. However, it can also lead to unwanted discussions. Developing a good (but of course also compliance-rules obeying) relationship to high-quality suppliers can be a strategic advantage and might warrant additional effort in the communication of the selection results.

14.2.6 Legal Perspective

The legal perspective of an AIoT initiative is often closely related to sourcing activities because customer/supplier relationships need a solid legal foundation. The following interview with Philipp Haas (head of the Expert Group for Digital and New Businesses at Bosch's legal department) provides some insights on the level perspective, building on the ACME Smart Shuttle example we introduced earlier (Fig. 14.19).

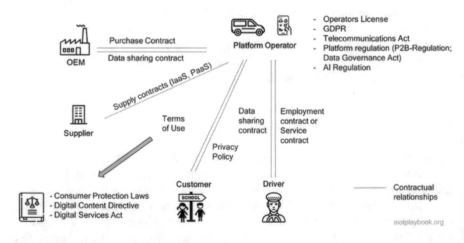

Fig. 14.19 Legal perspective – shuttle bus example

Dirk: *Thanks for joining us today. Tell us a little bit about what you do at Bosch. What's your role?*

Philipp: *I have been a consultant in the legal department of Bosch for 10 years now, and I am currently responsible for the central department for digital and new businesses. This includes consulting various smaller legal entities and central departments within Bosch. In addition, I'm also heading the expert team for IT law. We are also supporting the other colleagues in the legal department with respect to digital businesses.*

Dirk: *Thank you for supporting us with the Digital Playbook. When we started our discussions, we learned that the different legal aspects around AIoT are quite complex. That is why we said the best way to get a 360-degree view of the legal aspects would be to discuss this based on a realistic use case. So from a legal point of view, what are the key issues that we need to consider in our ACME Smart Shuttle example?*

Philipp: *I think the most important role is that of the platform operator because they sit in the middle of everything and offer the AIoT-enabled product. They have contractual relationships to many parties.*

Dirk: *Good point. Let's get started with the relationship between the platform operator and the OEM. What does the platform operator have to look out for from a legal perspective?*

Philipp: *In our scenario, the ACME Smart Shuttle is operating a fleet of shuttle buses, which need to be purchased or leased from the OEM. What is very important for our platform provider is that he's not only getting access to the vehicles, but that he is also having access to the data in the vehicles. Otherwise it will be more difficult to offer data-based services, which is a key assumption in this example. So there needs to be an additional agreement for the data generated by the vehicles. This means we need a data sharing contract. If the fleet is large enough, this could be an individually negotiated contract; alternatively, the platform provider has to agree to the standard offerings, which some OEMs already have out there. For example, BMW offers connected drive services, which include access to car data.*

Dirk: *Thanks. So that is our main supplier. What about the other suppliers, anything specific to look out for?*

Philipp: *Almost all IoT use cases today require a cloud provider, typically from the US or China. Cloud services are essential for the platform provider because they provide the infrastructure for running the software and the AI algorithms. Depending on the setting, you choose between software-as-a-service or infrastructure-as-a-service, if you need more control. Many of these cloud services are highly standardized today, and there will be little room for negotiating individual contracts. So selection of a cloud infrastructure player will not only be a technical choice but also requires you to look at costs and standard legal terms and conditions.*

Dirk: *And what about the counterpart to the cloud, the edge side of things. For example, in our Shuttle Bus example, we are assuming that there will be custom edge nodes embedded into the buses. What are the relevant aspects from a legal point of view with respect to the edge component provider?*

Philipp: *If the platform operator purchases devices that are responsible for the connectivity – for example, to his back-end – it might be necessary to have an agreement regarding the transport of the data. Such devices typically have a SIM card, either as a regular SIM or a built-in SIM card. It makes a difference who is responsible for the activation of this card. Therefore, if the device supplier is activating the card, it might be necessary that the supplier register as a telecommunication provider. The alternative would be, that the platform provider might have to conclude an additional contract with a responsible telecommunication provider directly. This might also be the case if the supplier is just delivering the hardware with a SIM card and the platform operator is responsible for activating the hardware (and the SIM). If the platform operator is responsible for activating the hardware, we have to examine his role. He then might become a telecommunication provider if he is responsible for the transport of data to his contract partners, but in our use case I do not think this will be the case for the platform provider.*

Dirk: *Talking about data in our Shuttle Bus scenario, one option that we have been discussing is for the bus operator to out-source the development and training of the AI algorithms. This would require the platform operator to make all the required data available to a third-party IT development firm. Are there any specifics that he has to look out for, in particular with respect to the ownership of his data?*

Philipp: *Yes, this is a very typical scenario. You are using the wording "his data", so the first question would be what exactly is "his data"? Does the data we are talking about truly belong to him? Legally there is no data ownership. If you're talking about data, there are two key aspects. The first key aspect is, are we talking about personal data? Because personal data within Europe are subject to the GDPR (General Data Protection Regulation), in addition to other international data protection laws. It typically means that you are only allowed to process the data – including handing it over to a third party for development – if you have a legal basis for that. The second key aspect for processing or transferring data are the relevant contracts. For example, the contracts that apply when receiving data might limit your ability to make these data available to a third party for further processing. So you're only allowed to transfer the data within these boundaries. If that is possible, usually there is no other legal protection for the data. In some very limited cases, data might also be protected by IP rights.*

Dirk: *In our scenario, the IT supplier of the ACME Smart Shuttle uses data from different sources, including data from the ACME Smart Shuttle, data from schools (e.g., school time tables), and data from third parties (e.g., traffic data). From these data, they derive new data via AI, e.g., bus schedules and routes. Does the AI and the new data created by the AI automatically belong to the ACME Smart Shuttle, because they are paying for it?*

Philipp: *No. It is highly recommended – I would say even absolutely necessary – to have a clear agreement with the IT supplier regarding the results that are created with the data. That is one of the topics I mentioned before, where it is legally not easy to determine who contributed to the results and who is the owner with respect to the results. That is why it is essential to have an explicit agreement on that. In joint development projects, you always have clauses regarding the rights to the development results. You also have clauses regarding software, so that, that is quite standard. In the newer contracts, we see clauses that explicitly refer to data, the right to data, maybe distinguishing between test data and productive data, and also with respect to work results that have been created using such data.*

Dirk: *I do understand the differentiation between software and data. But what about the trained AI models – are they data or software, from a legal point of view?*

Philipp: *An AI model will usually fall in the category of software. Software is defined in copyright law, and it is a program for computers that shows the computer what the next steps are. A trained AI model usually runs within a software environment. Maybe it is not a software on its own, it is just part of a software but also parts of software are considered as software under the copyright act. So I think it will be protected by copyright law, which means that it is possible to have an agreement on the usage rights and you can transfer that to the platform provider. And the platform provider will, of course try to do that because as you mentioned, he paid for it. However, this is not always possible because sometimes, if you have very large suppliers who argue that they are also using pre-existing works for their work results, it might be not easy to get all exclusive usage rights. There might be an individual agreement on who is allowed to do what with the work results.*

Dirk: *OK, let us assume we got all this sorted out, and we now have our platform up and running. What about our relationship to the end customer, the students of the school?*

Philipp: *I would say that is pretty straight forward. You offer your services most of the time via an app and for that app you need terms of use. We have standards that we are using for all different kinds of apps. And that platform provider has to comply with the relevant consumer protection laws that give very detailed requirements and that are renewed very often. In this year in Europe, we have some new consumer protection laws. You can also think about EULA's (End User License Agreements). You can use that in addition to the terms of use. So the terms of use cover the usage of the service itself, and the EULA is for the software. I do not think that it's necessary to use both.*

Dirk: *Another important group of stakeholders in our example are the drivers...*

Philipp: *In our example, the drivers are employed at platform operators. There might also be a service contract with them if they're independent, but then you have to make sure that they are truly independent and not "by-accident" employees, because this could cause major risks for the platform operator for example regarding tax law. The employment contract itself, I would say that is also quite standardized but we have this special case here that we need to have an agreement regarding the usage of the data from the shuttle buses. Because data that we get out of the vehicles could be contributed by the driver, it means that they are personnel related and that is why we need to have an agreement on the usage.*

This is legally not trivial because the platform operator has to obtain free and voluntary consent from his employee. I think in our use case, there is also a good justification for the platform provider because the usage of the data is an essential requirement for his business use case. He cannot operate the platform without that. So the request is absolutely reasonable.

Dirk: *Thanks. Anything else that we have to look out for from the perspective of the Shuttle Bus platform operator regarding legal aspects?*

Philipp: *We looked at the contractual relationships and I mentioned new legal developments regarding consumer protection laws. The same applies for digital business in general. There are many laws that either recently came into force or are still in development. I mentioned the telecommunications act that is currently revised on the European level. There are various legal drafts regarding platform regulation, and already existing platform regulations. The Data Governance Act will contain requirements if you want to share data via a platform. The Digital Content Directive has already been transformed into German law and such new regulations as of January 1, 2022. It makes various requirements for digital offers, which also includes software as a service or apps. For example, it contains an obligation to make regular security updates during the lifetime of the service. And on the horizon, we also see a regulation for AI. There is a first draft from the European Union. This is a very interesting regulation from a legal perspective. From the operator's perspective, it could lead to some new obligations, such as checking the data that he is using for the training of the AI models. According to the draft, the data have to be free of errors. There is an obligation to document the data usage. You have to document the results of the AI system so that you can track back exactly why a certain decision has been made by the AI. For nearly all AIoT products that are considered "high risk", this AI regulation will play a large role in the future.*

Dirk: *And do you see something similar coming up in USA and China as well?*

Philipp: *In the US, I recently read a statement from the US Department of Commerce regarding the AI Regulation, and it did not sound like they want to follow us. They seem to have a different approach and are looking with a skeptical eye on our regulation and do not think that it is helpful. So no, I don't expect a similar regulation from the US at the moment. In China, the situation is different. There are new security laws put into place and they also regulate the usage of the data. AI regulations are not for protecting the individual but more for protecting the interests of the government and the country. There will be a definition of categories for data that fall under these new security laws, but I read that vehicle data will be considered as one of the critical data categories. So I think in the future operating such a platform for China might be only possible within China.*

Dirk: *Last question. Looking at this from the perspective of the project manager, when in the lifetime of their project should they involve legal expertise? And what's the best way to actually embed this legal expertise in the project?*

Philipp: *Okay, this question is very simple to answer: As early as possible. Because there are many legal considerations and I would also say many traps.*

Dirk: *So depending on whether the operator operates from within a large organiza-*
tion or actually as a startup, how does he go about this? Does he really make
legal experts part of his team or how does he get access to this expertise?

Philipp: *This is a case by case decision. The legal counsel can become part of the*
project team, which has the advantage that he has deep knowledge about the
technical and business considerations of such an offering. For a startup it might
be too costly to involve external counsel as part of your project team and let them
participate in every discussion. You might take a leaner approach and discuss it
with the counsel and work out a plan at the beginning so that it is clear what has
to be considered. And then you can go ahead and have regular meetings, discus-
sions with the legal counsel, but not directly include him into every discussion.

Dirk: *Great. That was super interesting, thank you very much.*

Chapter 15
Rollout and Go-to-Market

Dirk Slama

Fig. 15.1 Agile AIoT grid

How to introduce our smart, connected products and solutions to their customers, be they external or internal, B2B or B2C? For solutions, this usually involves a dedicated rollout process, while for products the Go-to-Market is important (Fig. 15.1).

D. Slama (✉)
Ferdinand Steinbeis Institute, Berlin, Germany
e-mail: dirk.slama@bosch.com

© The Author(s) 2023
D. Slama et al. (eds.), *The Digital Playbook*,
https://doi.org/10.1007/978-3-030-88221-1_15

15.1 Smart, Connected Solutions: Rollout

Effective management of the rollout process for AIoT-enabled solutions is a key success criterion. How exactly this looks like will depend on many factors: is this for one site or one asset only, or is this for multiple sites with multiple assets? Is this for internal customers only, or for external customers? Will this require customizations for individual target sites?

We have already discussed several different examples representing different scenarios, e.g.,

- Single site, single asset: monitoring of particle collisions at the Large Hadron Collider
- Multi-site, multi asset: Rollout of a predictive maintenance solution for escalators in train stations
- Multi-site, multi asset with customization: Rollout of a predictive maintenance solution for different users of hydraulic components (requires customization of the AI for each customer)

The following describes a generalized process that could be suitable, for example, for a multi-site, multi asset situation.

During rollout preparation, a portfolio of all relevant assets and sites (e.g., train stations and escalators) must be created. This portfolio must be evaluated and prioritized. Based on this assessment, a project plan including rollout schedule and resource management must be created.

Rollout execution will then require a generalized plan which can be applied to each individual site. In this case, it includes site preparation (e.g., aligning with the train station's facility management, preparing for deployment), asset preparation (e.g., cordoning off the escalators, enabling access to the required internal parts), solution deployment (e.g., deploying an IoT gateway and an ultrasound sensor), testing the solution (e.g., simulating a problem with the escalator and checking if this is recognized by the solution), and finally transferring everything to operations.

Back to the portfolio level, the next area is performance and control. For example, this can monitor the rollout of the escalator monitoring solution and suggest corrections in case of inefficiencies. Finally, if this is a fixed set of sites/assets, the rollout project needs to be closed properly. This will include preparing measures for new assets being onboarded. For example, new escalators acquired in the future should also be equipped with the monitoring solutions (Fig. 15.2).

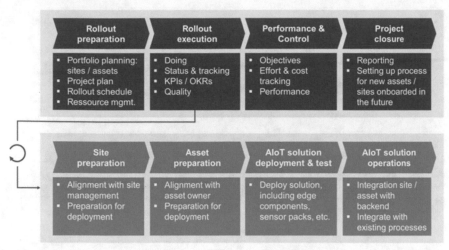

Fig. 15.2 Rollout of AIoT-enabled solutions

15.2 Smart, Connected Products: Go-to-Market

For the Digital OEM, a strong focus on the commercialization of new digital/physical offerings is key. For an incumbent, this needs to start with a look at existing sales and marketing processes, as well as the skills and networks of the existing team. How can this be applied to successfully market and sell new, digital/physical offering? And which changes might be required?

For digital/physical offerings, we often need a much closer alignment between the product development and the marketing/sales organization, since both marketing and sales functions need to be digital and built directly into the product. This is particularly true for any kind of digital subscription services or digitally managed physical-feature-on-demand offerings (Fig. 15.3).

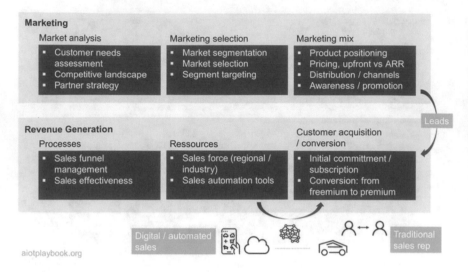

Fig. 15.3 Go-to-market

15.2.1 Example: Physical-Feature-on-Demand

If the new offerings include some kind of fleet/asset/feature-as-a-service element, it will be important that the sales organization be adopted accordingly. This will include many aspects, including sales commissions and incentives, sales processes, and customer engagement.

An interesting *on-demand* example is *seat-heating-on-demand*, as shown in Fig. 15.4. Traditionally, seat heating is sold as an add-on during the car sales process. Only if it is configured in the beginning will the car be equipped with it in the factory. The *on-demand* version assumes that all cars are equipped with the seat heating functionality. Customers can then use the car app to activate the feature on demand. The pricing for the feature could be dynamic, determined by an AI. In this example, responsibility for selling this feature would move from the sales rep, who is selling the car in the first place, to the team, which is responsible for demand generation for digital features. Another aspect is the change from a traditional, one-time payment via bank transfer to a subscription model based on micropayments.

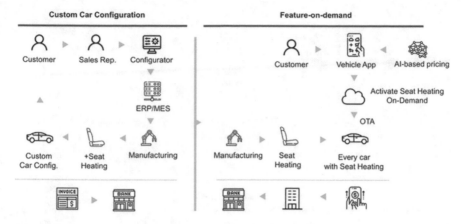

Fig. 15.4 Example: seat heating as physical-feature-on-demand

15.2.2   *Continuously Improve Commercialization*

One cannot expect to get the product/customer fit from the very beginning. So a clear focus on the continuous improvement of all relevant aspects of commercialization is required. In the example shown here, three key KPIs have been identified: number of signups, how many customers are actively using the freemium services, and how many are paying for premium services. There will always be a gap between these three KPIs, but of course the challenge is to drive them all up and minimize the gap. In order to do this, one will constantly have to monitor the customers along their customer journey. AIoT-generated insights can play a key role here, in addition to the standard digital analytics channels. The learning from the analysis must be applied for the continuous improvement of the offering and its commercialization: marketing and sales processes and campaigns can be adapted almost in real-time, especially if they are driven through digital channels. The digital product features can be adapted usually in a relatively short term, e.g., using Over-the-Air capabilities. Finally, even the physical product can be improved from generation to generation using the insights from the analysis of the customer journey. Managing this continuous improvement process effectively will be key to successfully scaling up an AIoT-enabled, digital/physical business (Fig. 15.5).

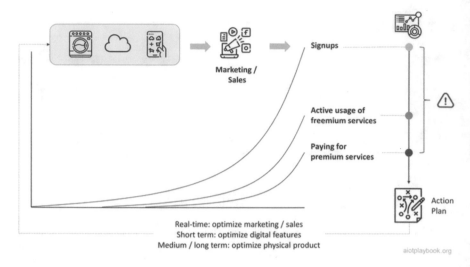

Fig. 15.5 AIoT commercialization & continuous improvement

Chapter 16
Operations

Dirk Slama

Providing efficient and effective service operations will be a key success factor for any AIoT-enabled product or solution. Depending on the specific nature of the system, service operations setup can potentially take very different shapes. For complex, industrial assets, service operations will most likely include direct customer interactions via a call center, as well as on-site maintenance or repair services. For mass-market consumer products, service operations will most likely be highly automated and provide only limited field services, if any. Most Field Service Management (FSM) organizations will be able to greatly benefit from AIoT-enabled features, which provide real-time access and advanced analytics of asset-related field data, or even support for predictive or preventive maintenance services.

Since the operations perspective will usually be quite different for the Digital OEM vs. the Digital Equipment Operator, this chapter will look at both perspectives.

16.1 Digital OEM (Fig. 16.1)

The operations perspective of the Digital OEM and his AIoT-enabled products will include a number of different elements. The sales organization will be responsible for supporting the new, digital-enabled features and services. The support organization must be able to handle the added product complexity. Finally, the DevOps organization must be able to continuously enhance and optimize the digital product offering.

D. Slama (✉)
Ferdinand Steinbeis Institute, Berlin, Germany
e-mail: dirk.slama@bosch.com

© The Author(s) 2023
D. Slama et al. (eds.), *The Digital Playbook*,
https://doi.org/10.1007/978-3-030-88221-1_16

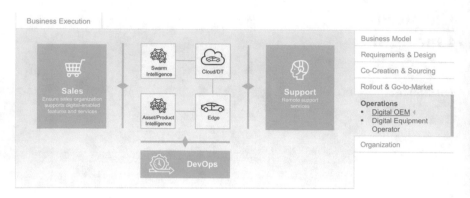

Fig. 16.1 Operations

16.1.1 Sales

Understanding digital transformation from a sales perspective is essential for its success. The Digital OEM is presented with many opportunities, which must be properly adopted by the sales organization.

AIoT will provide the sales and marketing organization with the opportunity to truly understand how customers are using the products in the field. Together with other data, e.g., from web analytics, CRM and social media, this will enable the sales and marketing organization to better target new and existing customers, e.g., for upselling newly available, digital-enabled features (Fig. 16.2).

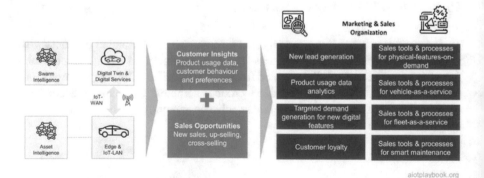

Fig. 16.2 AIoT-enabled sales organization

16.1.2   Support

Providing AIoT-enabled digital features can significantly increase a product's complexity. While it should be a core duty of the DevOps team to ensure the best possible user experience, there is a good chance that the new, digital features will cause additional customer requests to the support organization. There is nothing more frustrating for a customer buying a smart, connected product – let us say a vacuum robot – and then failing to get it to work, e.g., because of a pairing problem, or some other issue. Connectivity alone can be a source for many problems, which need to be addressed by the support organization. Especially for mass-market products, an efficient triage to manage the combination of internet FAQs, automated bots and potentially call center services will be important.

The support organization must also be prepared to deal with new, unexpected problems. For example, the use of AI in a smart, connected product might lead to problems that will initially be very hard to reproduce because the product is no longer following the deterministic logic encoded in the software (but rather is driven by an AI that is a black box in that regard).

Finally, the support organization should be supported with AIoT-enabled problem analytics and diagnostics. This will have to be provided by the DevOps team, which needs to focus not only on the product features but also on how to support the rest of the organization with AIoT-based features.

16.1.3   DevOps

While DevOps has the word *operations* in its name, the focus of the DevOps organization is usually on developing and operating smart, connected products. As discussed in the previous section, the focus of the DevOps team is usually on continuously improving the features of the product. However, one should not underestimate the importance of ensuring that the DevOps organization also supports the other parts of the operations side. In particular, the DevOps team will be responsible for providing sales, marketing, and support organizations with the required capabilities. Together, they need to identify which additional features – beyond the features important and visible to the end-user – will have to be built. The earlier example of *seat-heating-on-demand* applies here, where the DevOps team will not only have to build the feature itself but also implement dynamic pricing together with the sales team and build suitable in-app promotions in collaboration with the marketing team. Similarly, the DevOps team will be responsible for providing the support team with the required data, analytics reports and applications.

16.2   Digital Equipment Operator (Fig. 16.3)

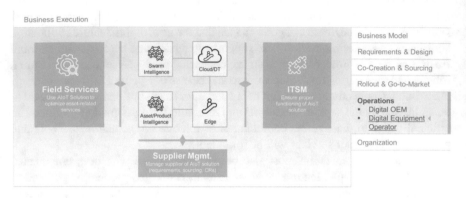

Fig. 16.3 Service operations

The Digital Equipment Operator will usually have a different perspective on the operations of the AIoT-enabled solution. This will most likely include field services related to assets in the field, IT Service Management related to the AIoT solution, and supplier management for the AIoT solution.

16.2.1   Field Service Management

Field service management (FSM) focuses on enterprise assets, e.g., operational equipment, machines and vehicles. FSM is described by Gartner [16] as a practice that *"includes the detection of a field service need (through remote monitoring or other means, inspection or a customer detecting a fault), field technician scheduling and optimization, dispatching, parts information delivery to the field, and process support of field technician interactions."*

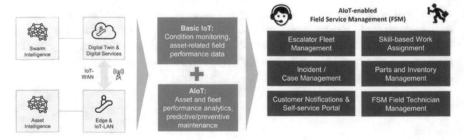

Fig. 16.4 AIoT & field service management

Figure 16.4 outlines how AIoT and FSM can play together. FSM can benefit from AIoT in a number of areas, including:

- Improved triage: Utilize AIoT to determine the severity and priority of asset-related incidents.
- Faster identification of required parts: Utilize AIoT for precise identification of assets and key parts deployed in the field.
- Inventory tracking: Utilize AIoT to create a precise and real-time inventory update.
- Initiation of automated intelligent dispatch events: Utilize AIoT to better prioritize incidents and to provide more information for problem resolution.
- Remote monitoring and diagnostics: Use real-time machine data for asset health and performance assessments.

All of this will only be possible if the AIoT project prepares the service operations organization accordingly. This will be one of the big challenges of the AIoT project management team. How to do this will greatly depend on a number of different factors, including:

- Is there already an existing organization responsible for FSM?
- If so, how is the organizational relationship between the IoT solution project and the existing FSM organization?
- If not, how far is the IoT solution project empowered to actually set up a new FSM organization to start operating after the start of production?
- Will the focus be mainly on operational FSM topics, or will it also include strategic topics such as Asset Performance Management (APM)?

16.2.2 IT Service Management

Another important dimension of AIoT Service Operations will be what is traditionally referred to as IT Service Management (ITSM). AIoT-ITSM will be responsible for ensuring the design, planning, delivery, operations, and management of all IT services related to the AIoT-enabled system. This means that AIoT-ITSM is not concerned with operating assets but rather enables the AIoT-features themselves. A well-established standard in the ITSM space is ITIL, the Information Technology Infrastructure Library. Without AIoT-ITSM, an AIoT system cannot be operated, which will be covered below.

ITIL defines five processes and four functions. The four functions are service desk, technical management, application management, and IT operations management. The five service operations processes are [17]:

- Access Management: grants authorized users the right to use a service; blocks any access request of non-authorized users to the service
- Event Management: captures, filters, and categorizes events to decide the appropriate actions to be taken. Events might or might not require an action.

- Incident Management: Incidents are events that have a negative impact on a service or its quality. Incident management helps restore IT service to a working state as quickly as possible.
- Problem Management: deals with identifying and addressing problems at their root. Multiple incidents can relate to the same problem.
- Request Fulfilment: responsible for acknowledging and processing service requests received from users. Usually, these are technical requests, not requests related to the functionality of business applications.

To manage all IT assets and other related data, ITIL foresees the use of a so-called configuration management database (CMDB) as the central repository for this kind of information. However, the complexity of introducing a CMDB should not be underestimated. Rouse [18] warns that CMDB projects often fail due to stale and unusable data. This is certainly an aspect that needs to be addressed, ideally by automating configuration data management as much as possible. Figure 16.5 provides an overview of how some key ITIL concepts can be applied to the AIoT perspective.

ITIL Areas / Processes	AIoT-ITSM Example
Service Design	
Service-Level Management	SLA for accuracy and response times of Machine Learning Cloud Services
Availability Management	Recovery procedure after failed FOTA update
Service Transition	
Knowledge Management	Procedure for configuring IP connectivity of IoT gateway
Release and Deployment Mgmt.	Upgrade of time series database server version
Service Testing and Validation	Testing of FOTA capabilities
Configuration Management System	Configuration of VPN for IoT gateways
Service Operation	
Event Management	Successful backup of IoT time series database
Incident Management	Asset / IoT gateway cannot connect to backend
Problem Management	Backend not available due to server downtime—asset fleet offline

aiotplaybook.org

Fig. 16.5 AIoT & IT service management

The architecture and organization for the supporting systems of the service operation will always be highly project-specific. However, the following discussion can provide some guidance regarding the architectural setup.

A key question is as follows: will there be separate AIoT-ITSM and FSM organizations, or will they be merged into one organization? While process-wise there might be similarities, the required skills will usually be very different. For example, the skills required to deal with the IP configuration of an IoT gateway or to keep a time series database running are very different than, for example, the skills required to analyze and repair the malfunction of an escalator. Consequently, the project must make a deliberate decision on how to organize AIoT-ITSM and FSM.

16.2.3 Option 1: Separate Systems

If it is decided that AIoT-ITSM and FSM will be two separate organizations, it can also make sense to run two separate support systems. As an example, a simplified monitoring solution for excavators is shown in Fig. 16.6, using some form of IoT gateway on the excavator. Both the FSM application and the AIoT-ITSM application have their own databases, receiving data from the gateway/TCU. The AIoT-ITSM solution uses some form of CMDB to store information related to the configuration items that make up the IoT solution (e.g., an inventory of gateways in the field, with related incidents). The FSM solution stores asset-related data, e.g., performance data from the hydraulics component of the excavator. Both solutions then have their dedicated and specialized staff, which supports their respective services.

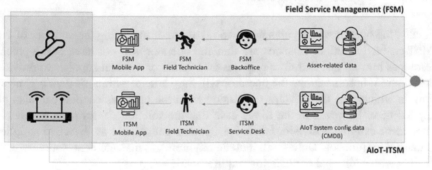

Fig. 16.6 Architecture: separate systems

16.2.4 Option 2: Integrated System

For strategic reasons, it can make sense to integrate AIoT-ITSM and FSM into the same organization, supported by an integrated system. In this case shown in Fig. 16.7, only one repository is used, which stores both asset-related and IoT enablement-related data. The back office supports all functions, and so is the field service. Of course, these are only two examples of a potential organizational setup; in reality, many other, potentially hybrid combinations could be possible. However, these examples serve the purpose of highlighting the issue and the choices an AIoT project manager must make.

Fig. 16.7 Architecture: integrated systems

16.2.5 Supplier Management

Finally, the operations side also needs to take care of managing the supplier of the AIoT-enabled solution. Chances are that the operator will not have development resources himself and therefore requires an internal or external supplier to provide the solution. In some cases, the operator will have a team within his own organization, in which case the DevOps discussion from the previous chapter can also be applied here. However, in the likely case that the solution comes from another – internal or external – division, then the operator must build an effective supplier management function. Duties will include requirements management, sourcing, and dealing with additional or changing requirements.

Take, for example, the railway operator example from the Introduction section. In this case, the railway operator acquired an AIoT-enabled solution for escalator monitoring. It is highly likely that this solution will be externally sourced, so supplier management becomes an integral part of the railway operator's organization. Ensuring the integration of the external escalator monitoring solution with the internal systems of the railway operator will be one key responsibility of this team.

Another interesting question will then be who will take on the responsibility for the IT service management of the escalator monitoring solution: will this be done in-house, or will the railway operator have a long-term support contract with the provider of the escalator monitoring solution? If this requires in-depth knowledge about other operational systems, then there is a good chance that at least parts of system operation (including the IT service management) will be in-house.

Chapter 17
Organization

Dirk Slama

The final perspective in the AIoT Business Execution discussion is on the organization, which needs to support the creation and operation of AIoT-enabled products or solutions. The organizational setup is a potential Achilles' heel: if this is not done properly, the entire initiative can be derailed. A number of different factors play a role here, from cultural aspects to proper alignment of the organizational structure with the key architectural elements of the product or solution. Owing to the large differences between Digital OEM and Digital Equipment operations from the organizational perspective, both will be discussed individually in this chapter.

17.1 Digital OEM (Fig. 17.1)

The product organization is ultimately responsible for delivering smart, connected products. Given the complexity of a typical AIoT product, as well as the different cultures that have to be brought together, building an efficient and effective AIoT product organization is not an easy task. This section of the AIoT framework discusses key challenges and proposes a specific setup that can be easily adapted to fit one's individual needs.

D. Slama (✉)
Ferdinand Steinbeis Institute, Berlin, Germany
e-mail: dirk.slama@bosch.com

© The Author(s) 2023
D. Slama et al. (eds.), *The Digital Playbook*,
https://doi.org/10.1007/978-3-030-88221-1_17

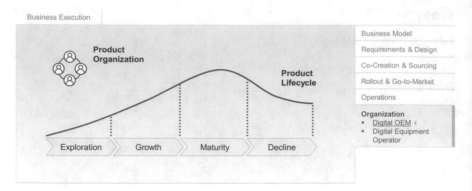

Fig. 17.1 Product organization

17.1.1 Product Organization

A successful product organization differs significantly from a project organization, as we will see in the following. In order to understand the differences, we will first look at the typical product lifecycle phases, before discussing how a product vs. a project organization typically evolves during this cycle.

17.1.2 Product Lifecycle Perspective

Particularly at the beginning of the new product journey, it is important to take a step back and look at the complete product lifecycle to be prepared for the road ahead. The lifecycle of most successful products can be described as follows:

- Exploration: During this phase, the initial product is developed, market reception is validated, initial customers are acquired, etc. Please note that the popular concept of a Minimum Viable Product (MVP) is more difficult to execute if physical product components are involved.
- Growth: The growth phase aims to expand reach, scale sales and continue to develop the product to stay competitive
- Maturity: In this phase, the focus is on customer retention and to sustain market share
- Decline: Finally, a strategic decision regarding strategic pivoting or phasing out has to be made; often the start for a next generation product

The interesting question now is: what must an organization look like to support a product through these different phases, and how does the organization itself have to evolve? (Fig. 17.2).

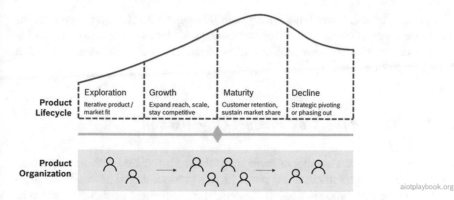

Fig. 17.2 Product lifecycle

17.1.3 Traditional Project Organization

In many incumbent enterprises, the development of a new product often starts as a project because from a controlling and administration point of view, setting up a project is more lightweight than establishing a new organizational unit. Since in the early stages it is often not known whether the product idea will be successful, this is quite understandable. If the initial MVP is promising, the project might be transferred to an internal accelerator. If the product shows the potential to scale from a sales point of view, a new line of business may eventually be created (Fig. 17.3).

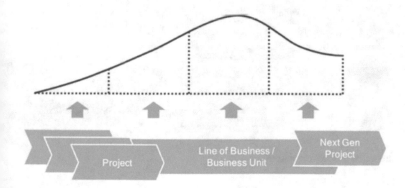

Fig. 17.3 Project organisations

In principle, there is nothing wrong with this approach. However, in practice it can cause severe problems. To better understand why, let us first quickly summarize the key differences between project and product. The table following provides a high-level comparison.

17.1.4   Toward the AIoT Product Organisation

In practice, there are a number of common problems associated with starting a new product with a project mindset and setup. First, a product should be developed from the beginning with product KPIs as the central measurement of success; customer satisfaction and customer adoption should be key KPIs from the very beginning. Typical project-centric KPIs such as development and go-to-market milestones (as well as cost) should be secondary.

Second, a typical project has a fixed start and end date, while a product needs to take a longer-term perspective. Especially in manufacturing-centric organizations, it is still a common assumption that at the end of the initial development project, the product is "ready" and can be handed over to a maintenance and operations team, eliminating the costs for expert developers. Such a transition will obviously cripple any long-term oriented, continuous advancement of the product (Fig. 17.4).

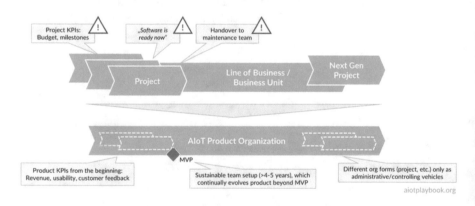

Fig. 17.4 Toward a real product organisation

Even if the product is initiated as a project in the early stages, it is key to remember the following:

- Use product-oriented KPIs from the beginning
- Implement a sustainable team setup with a 4 to 5-year perspective; this will continually evolve the product beyond the MVP
- View the project only as an administrative vehicle for initiating the product organization, but follow a product-oriented approach from the beginning

17.1.5   Organizational Culture

A key issue in most AIoT product organizations is the cultural differences typically found in heterogeneous teams that need to work together. Developers who are used to do frequent, cloud-enabled updates have a very different way of managing projects compared to manufacturing-centric engineers who know that after the SOP (Start-of-Production), any change to a physical asset after it has left the factory usually involves a costly and painful recall procedure. This *"Clash of two worlds"* within an AIoT product organization should not be underestimated. Actually, make this a *"Clash of three worlds"*: don not forget that the *"AI people"* usually also have a culture which is very different than the culture of the *"cloud/software people"*.

As shown in Fig. 17.5, different types of organizations typically have different types of cultures, even within the same company. Applying a one-size-fits all method to such a cultural mix will be difficult. This topic is discussed much more in-depth in the Agile AIoT section.

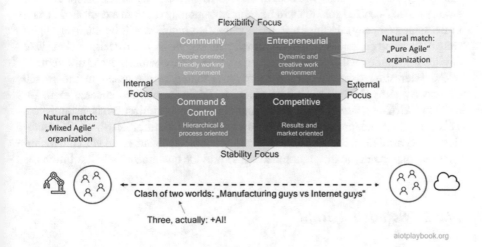

Fig. 17.5 Corporate cultures and Agile setup

17.2   Digital Equipment Operator (Fig. 17.6)

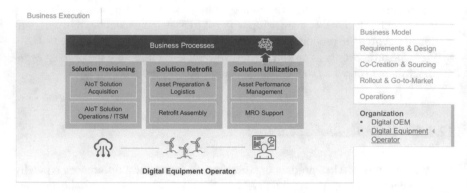

Fig. 17.6 DEO organization

The organization of the Digital Equipment Operator will usually be very different than that of the Digital OEM. The focus is much less on development but mostly on integrating the solution with the existing assets and the existing business processes. This will be discussed in the following.

17.2.1   Solution Provisioning

This part of the organization will heavily depend on the make vs. buy decision. In many cases, the AIoT-enabled solution will be sourced externally (or from a dedicated IT unit in the same company). In this case, the organization required will be relatively lightweight, focusing on requirements and provider management. If the decision is made to develop the solution with one's own resources, the picture obviously looks different.

An interesting example for solution provisioning is discussed in the Bosch Chassis Systems Control (CC) case study. In this example, a dedicated team for building AIoT-enabled solutions for manufacturing optimization is established. This central team supports manufacturing experts in the different Bosch CC factories. The central team has AI and analytics experts who then team up with the manufacturing experts in the different locations to provide customized AIoT solutions.

17.2.2   Solution Retrofit

Particularly in cases where new hardware (e.g. sensor packs) must be deployed to existing assets, solution retrofitting becomes a huge issue, and must be supported with the right organizational setup. Take, for example, the railway operator who wants to roll out the escalator monitoring solutions to thousands of escalators in different train stations around the country. For this rollout, a dedicated rollout/retro-fit organization will have to be established.

Depending on the complexity of the assets – how they are operationally utilized, and the scale of the rollout – this can be quite a significant organization. Of course, key questions include: for how long the rollout/solution retrofit organization must exist, and how the peak load during the initial rollout should be dealt with. Take, for example, an AIoT solution that needs to be retrofitted to all the trains in the railway operator's network. Each train might only get a couple of hours of extra mainte-nance time every year. This will be quite a challenge for the team responsible for applying the retrofit to a fleet of thousands of trains.

17.2.3   Solution Utilization

Ultimately, the utilization of the AIoT-enabled solution is indeed the aspect that is of most interest to the Digital Equipment Operator, as it is where the business ben-efit is generated. Depending on the nature of the solution, this can involve a dedi-cated organizational unit, or be supported by an existing unit. New, AIoT-enabled analytics features might feed into an existing MRO (maintenance, repair and opera-tions) organization. More advanced features, such as predictive maintenance, may already require some changes to the organizational setup because they will most likely have a more profound impact on the processes. For example, if the predictive maintenance solution actually predicts a potential failure, then the MRO organiza-tion must pick up this information and react to it. This process could be completely different than the traditional, reactive maintenance process.

If the AIoT solution offers a broader set of features to support Asset Performance Management (APM), then this will require a dedicated APM team to continually execute the performance optimizations.

Finally, if the AIoT solution feeds into other business processes, then business pro-cess re-engineering must be performed, and the new target processes must be supported by a suitable organizational setup. Take, for example, the AIoT-enabled flight path opti-mization system explained earlier. The introduction of such a system will have a signifi-cant impact on how the airline operates and touch many of its core processes.

Part IV
Technical Execution – AIoT Framework

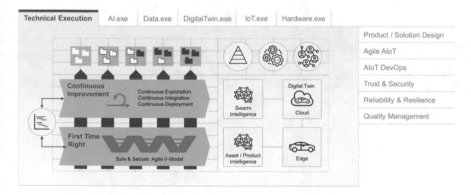

Fig. 1 Overview of AIoT Framework

Technical execution must ensure delivery of the AIoT-enabled product or solution in close alignment with the business execution. In the software world, this would usually be managed with an agile approach to ensure continuous value creation and improvement. However, in the AIoT world, we usually face a number of impediments that will prevent a pure agile setup. These impediments exist because of the typical complexity and heterogeneity of an AIoT system, including hardware, software, and AI development. In addition, an AIoT system usually includes components that have to be "first time right" because they cannot be changed after the Start of Production (especially hardware-based components or functionally relevant system components). Designing the system and the delivery organization in a way that maximizes those areas where continuous improvement can be applied while also efficiently supporting those areas where this is not possible is one of the key challenges of the technical execution.

The technical execution part of the *Digital Playbook* defines an AIoT Framework which looks at ways of supporting this. This starts with looking again at the data,

AI, IoT, Digital Twin and hardware perspective from the AIoT 101 chapter, but this time with the technical execution perspective ("*.exe").

In addition, this part provides a set of good practices and templates for the design of AIoT-enabled products and solutions, the implementation of an agile approach for AIoT (including the so-called "Agile V-Model"), AIoT DevOps (including cloud DevOps, MLops and DevOps for IoT), Trust & Security, Reliability & Resilience, Functional Safety, and Quality Management. Before going into detail, the following provides an overview of how all of these fit together, starting with the development life-cycle perspective.

Chapter 18
Development Life-Cycle Perspective

Dirk Slama

The development lifecycle of an AIoT-enabled product or solution usually includes a number of different sub-elements, which need to be brought together in a meaningful way. This chapter discusses this for both products and solutions.

18.1   Smart, Connected Products

Smart, connected products usually combine two types of features: physical and digital. The physical features are enabled by physical elements and mechanical mechanisms. The digital features are supported by sensors and actuators as the interface to the physical product, as well as edge and cloud-based components. Digital features can be realized as hardware, software or AI.

This means that the development life-cycle of a smart, connected product must include physical product development as well as manufacturing engineering. The development lifecycle of digital features focuses on DevOps for the edge components (including MLops for the AI deployed to the edge, DevOps for embedded and edge software, and embedded/edge hardware), as well as the cloud (including MLops for cloud-based AI and standard DevOps for cloud-based software).

All of this must be managed with a holistic Product Lifecycle Management approach. In most cases, this will require the integration of a number of different processes and platforms. For example, the development life cycle of the physical

D. Slama (✉)
Ferdinand Steinbeis Institute, Berlin, Germany
e-mail: dirk.slama@bosch.com

© The Author(s) 2023
D. Slama et al. (eds.), *The Digital Playbook*,
https://doi.org/10.1007/978-3-030-88221-1_18

features is traditionally supported by an engineering PLM platform, while software development is supported through a CI/CT/CD pipeline (Continuous Integration, Continuous Testing, and Continuous Deployment). For AI, these kinds of pipelines are different and not yet as sophisticated and mature as in the software world. The following will describe how such a holistic lifecycle can be supported (Fig. 18.1).

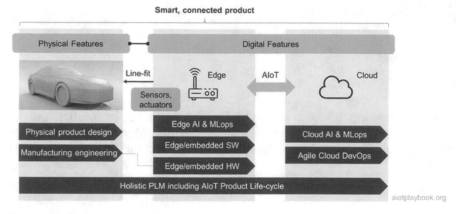

Fig. 18.1 Lifecycle – product perspective

Topics closely related to this include Cyber Physical Systems (CPS), as well as mechatronics. Mechatronics is an interdisciplinary engineering approach that focuses on the integration of mechanical, electronic and electrical engineering systems. The term CPS is sometimes used in the embedded world, sometimes with a similar meaning as IoT: integrate sensing and control as well as computation and networking into physical assets and infrastructure. Both concepts and the related development life-cycles can support smart, connected products.

18.2 Smart, Connected Solutions

For smart, connected solutions supporting the Digital Equipment Operator, the picture looks slightly different since physical product development is usually not within our scope. Sensors, actuators and edge nodes are usually deployed to existing assets in the field by using a retrofit approach. This means that the holistic lifecycle in this case does not include physical product design and manufacturing engineering. Other than this, it looks similar to the product perspective, expect that usually the required development pipelines will not be as sophisticated and highly automated as in the case of standardized product development (which typically invests more in these areas) (Fig. 18.2).

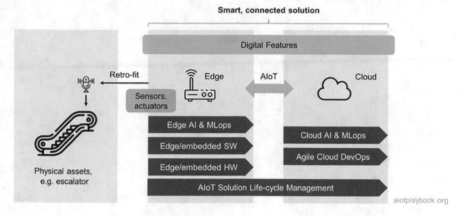

Fig. 18.2 Lifecycle – solution perspective

Chapter 19
Designing Smart Connected Products and Solutions with AIoT

Dirk Slama

An important element in the development lifecycle is the end-to-end design of the product or solution. The design section will provide a set of detailed templates that can be used here. These templates support the key viewpoints developed by the Digital Playbook: Business Viewpoint, UX Viewpoint, Data/Functional Viewpoint, and Implementation Viewpoint. These design viewpoints must be aligned with the agile product development perspective, in particular the story map as the top-level work breakdown. They will have to be updated frequently to reflect any learning from the implementation sprints. This means that they can only have a level of detail that permits them to do this (Fig. 19.1).

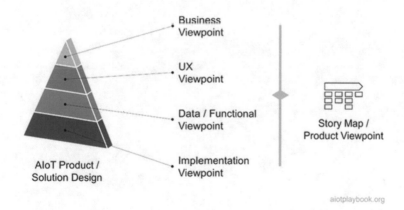

Fig. 19.1 AIoT design viewpoints

D. Slama (✉)
Ferdinand Steinbeis Institute, Berlin, Germany
e-mail: dirk.slama@bosch.com

© The Author(s) 2023
D. Slama et al. (eds.), *The Digital Playbook*,
https://doi.org/10.1007/978-3-030-88221-1_19

Chapter 20
AIoT Pipelines

Dirk Slama

Pipelines have become an important concept in many development organizations, especially from a DevOps perspective. This chapter introduces the concept of AIoT pipelines and discusses pipeline aggregations.

20.1 Definition

There are a number of different definitions for the pipeline concept. On the technical level, a good example is the popular development support tool git, which provides a set of tools to allow flexible creation of pipelines to automate the continuous integration process. On the methodological level, for example, the Scaled Agile Framework (SAFe) introduces the concept of Continuous Delivery Pipelines (CDP) as the automation workflows and activities required to move a new piece of functionality from ideation to release. A SAFe pipeline includes Continuous Exploration (CE), Continuous Integration (CI), Continuous Deployment (CD), and Release on Demand. This makes sense in principle.

The Digital Playbook is also based on the concept of pipelines. An AIoT pipeline helps move a new functionality through the cycle from ideation and design to release, usually in a cyclic approach, meaning that the released functionality can enter the same pipeline at the beginning to be updated in a subsequent release. The assumption is that AIOT pipelines are usually bound to a particular AIoT technical platform, e.g., edge AI, edge SW, cloud AI, cloud SW, smartphone apps, etc. Each AIoT pipeline usually has an associated pipeline team with skills specific to the pipeline and the target platform (Fig. 20.1).

D. Slama (✉)
Ferdinand Steinbeis Institute, Berlin, Germany
e-mail: dirk.slama@bosch.com

© The Author(s) 2023
D. Slama et al. (eds.), *The Digital Playbook*,
https://doi.org/10.1007/978-3-030-88221-1_20

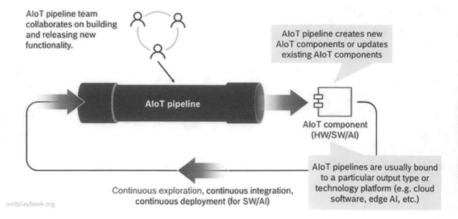

Fig. 20.1 AIoT pipeline – definition

20.2   Pipeline Aggregations

Due to the complexity of many AIoT initiatives, it can make sense to logically aggregate pipelines. This is something that many technical tools with built-in pipeline support such as git are providing out of the box. From the point of view of the target platform, the aggregation concept also makes sense. Take, for example, an edge pipeline that aggregates edge AI components, edge software components, and potentially even custom edge hardware into a higher-level edge component. On the organizational level, this can mean that a higher-level pipeline organization aggregates a number of pipeline teams. For example, the edge pipeline team consists of an edge AI and an edge software team.

This way of looking at an organization can be very helpful to manage complexity. It is important to note that careful alignment of the technical and organizational perspectives is required. Usually, it is best to create a 1:1 mapping between technical pipelines, target platforms and pipeline teams.

Figure 20.2 shows an edge pipeline that aggregates three pipelines, namely edge AI, edge HW and edge SW. The combined output of the three lower-level pipelines is combined into integrated edge components.

Fig. 20.2 AIoT pipelines aggregates

20.3   AIoT Pipelines & Feature-Driven Development

Technical pipelines are useful for managing and – at least partially – automating the creation of new functionalities within a single technology platform. However, many functional features in an AIoT system will require support from components on a number of different platforms. Take, for example, the function to activate a vacuum robot via the smartphone. This feature will require components on the smartphone, the cloud and the robot itself. Each of these platforms is managed by an individual pipeline. It is now important to orchestrate the development of the new feature across the different pipelines involved. This is best done by assigning new features to feature teams, which work across pipelines and pipeline teams. There are a number of different ways this can be done, e.g., by making the pipeline teams the permanent home of technology experts in a particular domain and then creating virtual team structures for the feature teams that get the required experts from the technical pipelines teams assigned for the duration of the development of the particular feature. Another approach can be to permanently establish the feature teams and look at the technical pipeline teams more as a loose grouping. Unfortunately, different technology stacks and cross-technology features tend to require dealing with some kind of organizational matrix structure, which must be addressed one way or another. There are some examples of how other organizations are looking at this, e.g., the famous Spotify model. *The Digital Playbook* does not make any assumptions about how this is addressed in detail but recommends the combination of pipelines/pipelines teams on the one hand, and features/features teams on the other (Fig. 20.3).

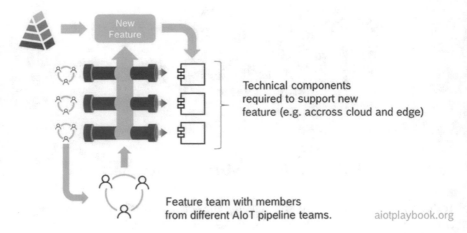

Fig. 20.3 AIoT features

Jan Bosch is Professor at Chalmers University and Director of the Software Center: *There are two different ways in which you're going to organize. In the component-based organizational model, you have the overall system architecture and assign teams to the different components and subsystems. The alternative model is a feature teams model; you have teams pick work items from the backlog. That team can then touch any component in the system and make all the changes they need to make to deliver their features. That is, in general, my preferred approach, but it is an important caveat. The companies that do this in an embedded systems context are associating the required skills typically with work items in the backlog. They say whatever team picks this up has to have at least these skills to deliver on this feature successfully. So it is not that any team can pick any work item.*

20.4 Holistic AIoT DevOps

The pipeline concept must be closely aligned with DevOps. DevOps is a well-established set of practices that combine software development and IT operations. In more traditional organizations, these two functions used to be in different silos, which often caused severe problems and inefficiencies. DevOps focuses on removing these frictions between development and operations teams by ensuring that developer and operations experts are working in close alignment across the entire software development lifecycle, from coding to testing to deployment.

An AIoT initiative will have to look at DevOps beyond the more or less well-established DevOps for software. One reason is that AI development usually requires a different DevOps approach and organization. This is usually referred to as MLops. Another reason is that the highly distributed nature of an AIoT system usually requires that concepts such as Over the Air Updates be included, which is another complexity usually not found in cloud-centric DevOps organizations. All of these aspects will be addressed in the AIoT DevOps section in more detail.

In addition to the DevOps focused on continuous delivery of new features and functionalities, an AIoT organization will usually also need to look explicitly at security and potentially functional safety, as well as reliability and resilience. These different aspects will have to be examined through the cloud and edge software perspective, as well as the AI perspective. *The Digital Playbook* builds on existing concepts such as DevSecOps (an extension of DevOps to also cover security) to address these issues specifically from an AIoT point of view (Fig. 20.4).

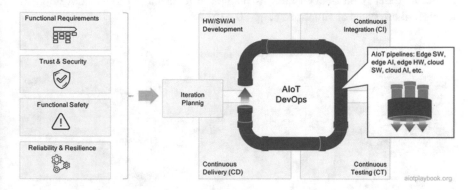

Fig. 20.4 AIoT pipelines + DevOps

20.5 Managing Different Speeds of Development

One of the biggest challenges in most AIoT projects is managing the different speeds of development that can usually be found. For example, hardware and manufacturing-related topics usually move much slower (i.e., months) than software or AI development (weeks). In some cases, one might even have to deal with elements that change on a daily basis, e.g., automatically retrained AI models. To address this, one must carefully consider the organizational setup. Often, it can make sense to allow these different topics to evolve at their own speed, e.g., by allowing a different sprint regime for different pipelines that produce AIoT artifacts and components at different speeds. An overview is given in the figure following. Please note that there is often no straight-forward answer for dealing with AIoT elements that require either very long or very short iterations. For example, for very slow moving elements, one can choose very long sprints. Alternatively, one can have all teams work with a similar spring cadence but allow the slower moving topics to deliver non-deployable artifacts, e.g., updated planning and design documents, etc. Similarly, for very fast moving elements the strict sprint cadence might be too rigid, so it could be better to allow them to be worked on and released ad hoc. For example, like automatically retrained AI models, this makes perfect sense since for an automated process no sprint planning seems required.

However, there is a key prerequisite for this to work: dependencies between artefacts and components from the different AIoT pipelines have to be carefully managed from a dependency point of view. In general, it is OK for fast moving artefacts
to depend on slower moving artefacts, but not the other way around – otherwise the
evolution of the fast moving artefacts will have a negative impact on the slower
moving artefacts. These dependencies can be of a technical nature (e.g., call dependencies between software components, or deployment dependencies between hardware and software) or of a more organizational nature (e.g., procurement decisions).
The technical dependencies and how to deal with them will be discussed in more
detail in the Data/Functional Viewpoint of the Product/Solution Design. Finally, the
Agile V-Model is introduced later as an option to manage product development
teams in these types of situations (Fig. 20.5).

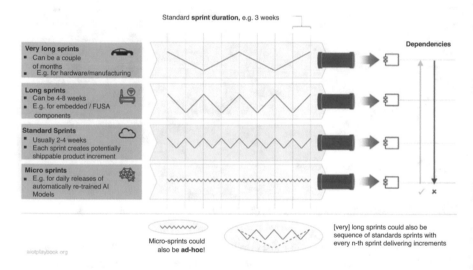

Fig. 20.5 Managing different speeds of development

Jan Bosch from Chalmers University and the Software Center: *This is a key
question: How do you do a release? There are companies in the earliest development stage that do heartbeat-based releases; every component releases every
third or every fourth week at the end of the agile sprints. You release all the new
versions of the components simultaneously, so that is one way. However, this
requires a high level of coordination between the different groups who are building different subsystems in different parts of the system. This is why many companies aim to reach a state where continuous integration and testing of the overall
system is so advanced that any of the components in the system can release at any
point in time, as long as they have passed the test cases. Then, the teams can start
to operate on different heartbeats. Some of the leading cloud companies are now
releasing multiple times a day. This should also be the goal for an AIoT system:
frequent releases, early validation, less focus on dependency management between
different teams.*

Chapter 21
AIoT.exe

Dirk Slama

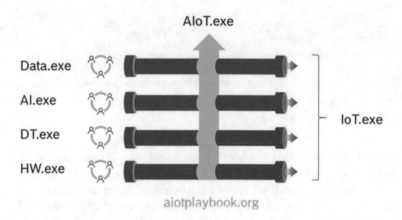

Fig. 21.1 AIoT.exe

The starting point of the discussion on technical execution following will be a deep dive into our topics from the AIoT 101 section, namely, AI, Data, Digital Twin, IoT, and Hardware: the key ingredients of many AIoT products and solutions. Each topic will be specifically looked at from the execution perspective (hence the play with "*.exe"), with a focus on both technology and organization. For each topic, we will also discuss how the technical pipeline and pipeline organization should be addressed and how it can all be integrated (mainly through the IoT perspective) (Fig. 21.1).

D. Slama (✉)
Ferdinand Steinbeis Institute, Berlin, Germany
e-mail: dirk.slama@bosch.com

 This will then be followed by a more detailed discussion on AIoT product/solution design, Agile AIoT, AIoT DevOps, and finally Trust & Security, Functional Safety, Reliability & Resilience and Quality Management.

21.1   AI.exe (Fig. 21.2)

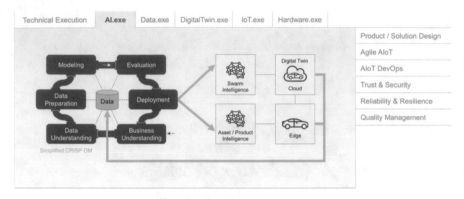

Fig. 21.2 Ignite AIoT – artificial intelligence

Naturally, AI plays a central role in every AIoT initiative. If this is not the case, then it is maybe IoT – but not AIoT. In order to get the AI part right, the Digital Playbook proposes to start with the definition of the AI-enabled value proposition in the context of the larger IoT system. Next, the AI approach should be fleshed out in more detail. Before starting the implementation, one will have to also address skills, resources and organizational aspects. Next, data acquisition and AI platform selection are on the agenda before actually designing and testing the model and then building and integrating the AI Microservices. Establishing MLops is another key prerequisite for enabling an agile approach, which should include PoC, MVP and continuous AI improvements.

21.1.1   Understanding the Bigger Picture

Many AIoT initiatives initially only have a vague idea about the use cases and how they can be supported by AI. It is important that this is clarified in the early stages. The team must identify and flesh out the key use cases (including KPIs) and how they are supported by AIoT. Next, one should identify what kind of analysis or forecasting is required to support these KPIs. Based on this, potential sensors can be identified to serve as the main data source. In addition, the AIoT system architecture must be defined. Both will have implications for the type of AI/ML that can be applied (Fig. 21.3).

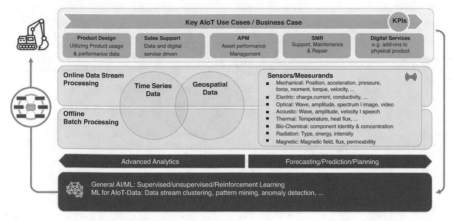

Fig. 21.3 AI value proposition and IoT

21.1.2 The AIoT Magic Triangle

The AIoT Magic Triangle describes the three main driving forces of a typical AIoT solution:

- IoT Sensors & data sources: What sensors can be used, taking physical constraints, cost and availability into consideration? What does this mean for the type of sensor data/measurements which will be available? What other data sources can be accessed? And how can relevant data sets be created?
- AIoT system architecture: How does the overall architecture look like, e.g. how to distributed data and processing logic between cloud and edge? What kind of data management and AI processing infrastructure can be used?
- AI algorithm: Finally, which AI method/algorithm can be used, based on the available data and selected system architecture?

The AIoT magic triangle also looks at the main factors that influence these three important factors:

- Business requirements/KPIs, e.g., required classification accuracy
- UX requirements, e.g., expected response times
- Technical/physical constraints, e.g., bandwidth and latency (Fig. 21.4)

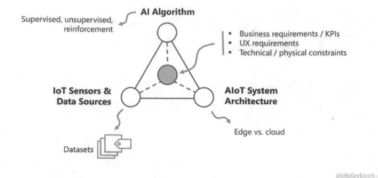

Fig. 21.4 The AIoT magic triangle

The AIoT magic triangle is definitely important for anybody working on the AIoT short tail (i.e., products), where there are different options for defining any of the tree elements of the triangle. For projects focusing on the AIoT long tail, the triangle might be less relevant – simply because for AIoT long tail scenarios, the available sensor and data sources are often predefined, as is the architecture into which the new solutions have to fit. Keep in mind that the AIoT long tail usually involves multiple, lower-impact AIoT solutions that share a common platform or environment, so freedom of choice might be limited.

21.1.3 Managing the AIoT Magic Triangle

As a product/project manager, managing the AIoT magic triangle can be very challenging. The problem is that the three main elements have very different lifecycle requirements in terms of stability and changeability:

- The IoT sensor design/selection must be frozen earlier in the lifecycle, since the sensor nodes will have to be sourced/manufactured/assembled – which means potentially long lead times
- The AIoT System Architecture must usually also be frozen some time later, since a stable platform will be required at some point in time to support development and productization
- The AI Method will also have to be fixed at some point in time, while the actual AI model is likely to continuously change and evolve. Therefore, it is vital that the AIoT System Architecture supports remote monitoring and updates of AI models deployed to assets in the field

Figure 21.5 shows the typical evolution of the AIoT magic triangle in the time leading up to the launch of the system (including the potential Start of Production of the required hardware).

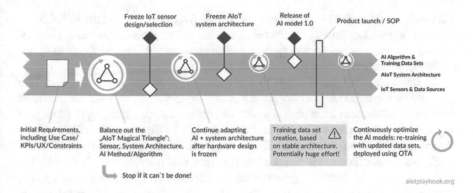

Fig. 21.5 AIoT magic triangle evolution

Especially in the early phase of an AIoT project, it is important that all three angles of the AIoT magic triangle are tried out and brought together. A Proof-of-Concept or even a more thorough pilot project should be executed successfully before the next stages are addressed, where the elements of the magic triangle are frozen from a design spec point of view, step by step.

21.1.4   First: Project Blueprint

Establishing a solid project blueprint as early as possible in the project will help align all stakeholders and ensure that all are working toward a common goal. The project blueprint should include an initial system design, as well as a strategy for training data acquisition. A proof-of-concept will help validate the project blueprint.

Proof-of-Concept
In the early stages of the evaluation, it is common to implement a Proof-of-Concept (PoC). The PoC should provide evidence that the chosen AIoT system design is technically feasible and supports the business goals. This PoC is not to be confused with the MVP (Minimal Viable Product). For an AIoT solution or product, the PoC must identify the most suitable combination of sensors and data sources, AI algorithms, and AIoT system architecture. Initially, the PoC will usually rely on very restricted data sets for initial model training and testing. These initial data sets will be acquired through the selected sensors and data sources in a lab setting. Once the team is happy that it has found a good system design, more elaborate data sets can be acquired through additional lab test scenarios or even initial field tests.

Initial System Design
After the PoC is completed successfully, the resulting system architecture should be documented and communicated with all relevant stakeholders. The system architecture must cover all three aspects of the AIoT magic triangle: sensors and data selection, AIoT architecture, and AI algorithm. As the project goes from PoC to MVP, all

the assumptions have to be validated and frozen over time, so that the initial MVP can be released. Depending on the requirements of the project (first-time-right vs. continuous improvement), the system architecture might change again after the release of the MVP.

It should be noted that changes to a system design always come at a cost. This cost will be higher the further the project is advanced. Changing a sensor spec after procurement contracts have been signed will come at a cost. Changing the design of any hardware component after the launch of the MVP will cause issues, potentially forcing existing customers to upgrade at extra cost. This is why a well-validated and stable system architecture is worth a lot. If continuous improvement is an essential part of the business plan, then the system architecture will have to be designed to support this. For example, by providing means for monitoring AI model performance in the field, allowing for continuous model retraining and redeployment, and so on.

Define Strategy for Training Data Acquisition and Testing

In many AI projects, the acquisition of data for model training and testing is one of the most critical – and probably one of the most costly – project functions. This is why it is important to define the strategy for training data acquisition early on. There will usually be a strong dependency between system design and training data acquisition:

- Training data acquisition will rely on the system architecture, e.g., sensor selection. The same sensor, which is defined by the system architecture, will also have to be used for the acquisition of the training data.
- The system architecture will have to support training data acquisition. Ideally, the systems used for training data acquisition should be the same system, which is later put into production. Once the system is launched, the production system can often be used to acquire even more data for training and testing.

Training data acquisition usually evolves alongside the system design – both are going hand in hand. In the early stages, the PoC environment is used to generate basic training data in a simple lab setup. In later stages, more mature system prototypes are deployed in the field, where they can generate even better and more realistic training data, covering an increasing number of real-world cases. Finally, if feasible, the production system can generate even more data from an entire production fleet.

Advanced organizations are using the so-called "shadow mode" to test model improvements in production. In this mode, the new ML model is deployed alongside the production model. Both models are given the same data. The outputs of the new model are recorded but not actively used by the production system. This is a safe way of testing new models against real-world data, without exposing the production system to untested functionality. Again, methods such as the "shadow mode" must be supported by the system design, which is why all of this must go hand in hand.

21.1.5   Second: Freeze IoT Sensor Selection

The selection of suitable IoT Sensors can be a complex task, including business, functional and technical considerations. Especially in the early phase of the project, the sensor selection process will have to be closely aligned with the other two elements of the AIoT magic triangle to allow room for experimentation. The following summarizes some of the factors that must be weighted for sensor selection, before making the final decision:

- Functional feasibility: does the sensor deliver the right data?
- Response speed: does it capture time-sensitive events at the right speed?
- Sensing range: does it cover the required sensing range?
- Repetition accuracy: does it tread similar events equally?
- Adaptability: can the sensor be configured as required, are all required interfaces openly accessible?
- Form factor: Size, shape, mounting type
- Suitability for target environment: ruggedness, protection class, temperature sensitivity
- Power supply: voltage range, power consumption, electrical connection
- Cost: What is the cost for sensor acquisition? What about additional operations costs (direct and indirect)?

Of course, sensor selection cannot be performed in isolation. Especially in the early phase, it is important that sensor candidates be tested in combination with potential AI methods. However, once the team is convinced on the PoC-level (Proof of Concept) that a specific combination of sensors, AI architecture and AI method is working, the decision for the sensor is the first one that must be frozen, since the acquisition of the sensors will have the longest lead time. Additionally, once this decision is fixed, it will be very difficult to change. For more details on the IoT and sensors, refer to the AIoT 101 and IoT.exe discussion.

21.1.6   Third: Freeze AIoT System Architecture

The acquisition of an AI platform is not only a technical decision but also encompasses strategic aspects (cloud vs. on premises), sourcing, and procurement. The latter should not be underestimated, especially in larger companies. The often lengthy decision-making processes of technology acquisition/procurement processes can potentially derail an otherwise well planned project schedule.

However, what actually constitutes an AI system architecture? Some key elements are as follows:

- Distributed system architecture: how much processing should be done on the edge, how much in the cloud? How are AI models distributed to the edge, e.g., via OTA? How can AI model performance be monitored at the edge? This is discussed in depth in the AIoT 101, as well as the data/functional viewpoint of the AIoT Product/Solution Design.

- AI system architecture: How is model training and testing organized? How is MLops supported?
- Data pipeline: How are data ingestion, storage, transformation and preparation managed? This is discussed in the Data.exe part.
- AI platform: Finally, should a dedicated AI platform be acquired, which supports collaboration between different stakeholders? This is discussed at the end of this chapter.

21.1.7 Fourth: Acquisition of Training Data

Potentially one of the most resource intensive tasks of an AIoT project is the acquisition of the training data. This is usually an ongoing effort, which starts in the early project phase. Depending on the product category, this task will then either go on until the product design freeze ("first-time-right"), or even continue as an ongoing activity (continuous model improvements). In the context of AIoT, we can identify a number of different product categories. Category I is what we are calling mechanical or electro-mechanical products with no intelligence on board. Category II includes software-defined products where the intelligence is encoded in hand-coded rules or software algorithms. Category III are "first-time-right" products, which cannot be changed or updated after manufacturing. For example, a battery-operated fire alarm might use embedded AI for smoke analysis and fire detection. However, since it is a battery-operated and lightweight product, it does not contain any connectivity, which would be the prerequisite for later product updates, e.g., via OTA. Category IV are connected test fleets. These test fleets are usually used to support generation of additional test data, as well as validation of the results of the model training. A category III product can be created using a category IV test fleet. For example, a manufacturer of fire alarms might produce a test fleet of dozens or even hundreds of fire alarm test systems equipped with connectivity for testing purposes. This test fleet is then used to help finalizing the "first-time-right" version of the fire alarm, which is mass produced without connectivity. Of course, category IV test fleets can also be the starting point for developing an AI which then serves as the starting point for moving into a production environment with connected assets or products in the field. Such a category V system will use the connectivity of the entire fleet to continuously improve the AI and re-deploy updated models using OTA. Such a self-supervised fleet of smart, connected products is the ideal approach. However, due to technical constraints (e.g., battery lifetime) or cost considerations this might not always be possible.

This approach of classifying AIoT product categories was introduced by Marcus Schuster, who heads the embedded AI project at Bosch. It is a helpful tool to discuss requirements and manage expectations of stakeholders from different product categories. The following will look in more detail at two examples (Fig. 21.6).

Categories

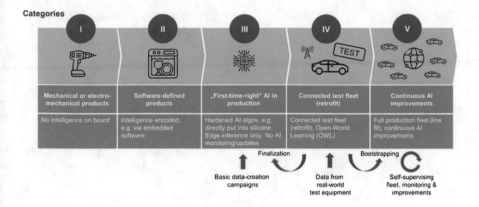

Fig. 21.6 AI product categories

Example 1: "First-Time-Right" Fire Alarm

The first example we want to look at is a fire alarm, e.g., used in residential or commercial buildings. A key part of the fire alarm will be a smoke detector. Since smoke detectors usually have to be applied at different parts of the ceiling, one cannot always assume that a power line or even internet connectivity will be available. Especially if they are battery operated, wireless connectivity usually is also not an option, because this would consume too much energy. This means that any AI-enabled smoke detection algorithm will have to be "first-time-right, and implemented on a low-power embedded platform. Sensors used for smoke detection usually include photoelectric and ionization sensors.

In this example, the first product iteration is developed as a proof-of-concept, which helps validate all the assumptions which must be made according to the AIoT magic triangle: sensor selection, distribution architecture, and AI model selection. Once this is stabilized, a data campaign is executed which uses connected smoke sensors in a test lab to create data sets for model training, covering as many different situations as possible. For example, different scenarios covered include real smoke coming from different sources (real fires, or canned smoke detector tester spray), nuisance smoke (e.g., from cooking or smoking), as well as no smoke (ambient).

The data sets from this data campaign are then validated and organized as the foundation for creating the final product, where the training AI algorithm is then put into or onto silicone e.g., using TinyML and an embedded platform, or even by creating a custom ASIC (application-specific integrated circuit). This standardized, "first-time-right" hardware is then embedded into the mass-manufactured smoke detectors. This means that after the Start of Production (SOP), no more changes to the model will be possible, at least not for the current product generation (Fig. 21.7).

Fig. 21.7 Example: "first-time-right" AIoT product (fire alarm)

Example 2: Continuous Improvement of Driver Assistance Systems

The second example is the development of a driver assistance systems, e.g., to support highly automated driving. Usually, such systems and the situations they have to be able to deal with are an order of magnitude more complex than those of a basic, first-time-right type of product.

Development of the initial models can be well supported by a simulation environment. For example, the simulation environment can simulate different traffic situations, which the driver assistance system will have to be able to handle. For this purpose, the AI is trained in the simulator.

As a next step, a test fleet is created. This can be, for example, a fleet of normal cars, which undergo a retrofit with the required sensors and test equipment. Usually, the vehicles in the test fleet are connected, so that test data can be extracted, and updates can be applied.

Once the system has reached a sufficient level of reliability, it will become part of a production system. From this moment onwards, it will have to perform under real-world conditions. Since a production system usually has many more individual vehicles than a test fleet, the amount of data which can now be captured is enormous. The challenge now is to extract the relevant data segments from this huge data stream which are most relevant for enhancing the model. This can be done, for example, by selecting specific "scenes" from the fleet data which represent particularly relevant real-world situations, which the model has not yet been trained on. A famous case here is the "white truck crossing a road making a U-turn on a bright, sunny day", since such a scenario has once lead to a fatal accident with a autopilot.

When comparing the "first-time-right" approach with the continuous improvement approach, it is important to notice that the choice of the approach has a fundamental impact on the entire product design, and how it evolves in the long term. A first-time-right fire alarm is a much more basic product than a vehicle autopilot. The former can be trained using a data campaign which probably takes a couple of weeks, while the latter takes an entire product organization with thousands of AI and ML experts and data engineers, millions or cars on the road, and billions of test miles driven. But then also the value creation is hugely different here. This is why it is important for a product manager to understand the nature of this product, and which approach to choose (Fig. 21.8).

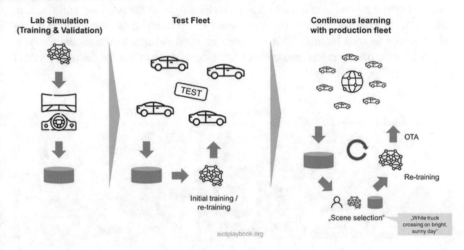

Fig. 21.8 Example: continuous improvement of AI models (driver assistance)

The AIoT Data Loop

Getting feedback from the performance of the products in the field and applying this feedback to improve the AI models is key for ensuring that products are perfected over time, and that the models adapt to any potential changes in the environment. For connected products, the updated models can be re-deployed via OTA. For unconnected products, the learning can be applied to the next product generation.

The problem with many AIoT-enabled systems is: how to identify areas for improvement? With physical products used in the field, this can be tricky. Ideally, the edge-based model monitoring will automatically filter out all standard data, and only report "interesting" cases to the backend for further processing. But how can the system decide which cases are interesting? For this, on usually need to find an ingenious approach which often will not be obvious in the first place.

For example, for automated driving, the team could deploy an AI running in so-called shadow mode. This means the human driver is controlling the car, and the AI is running in parallel, making its own decisions but without actually using them to control the car. Every time the AI makes a decision different from the one of the human driver, this could be of interest. Or, let us take our vacuum robot example. The robot could try to capture situations which indicate sub-optimal product performance, e.g., the vacuum being stuck, or even being manually lifted by the home-owner. Another example is leakage detection for pneumatic systems, using sound pattern analysis. Every time the on-site technician is not happy with the system's recommendations, he could make this known to the system, which in turn would capture the relevant data and mark it for further analysis in the back-office.

The processing of the monitoring data which has been identified as relevant will often be a manual or at least semi-manual process. Domain experts will analyze the data and create new scenarios, which need to be taught to the AI. This will result in extensions to existing data sets (or even new data sets), and new labels which represent the new lessons learned. This will then be used as input to the model re-training. After this, the re-trained models can be re-deployed or used for the next product generation (Fig. 21.9).

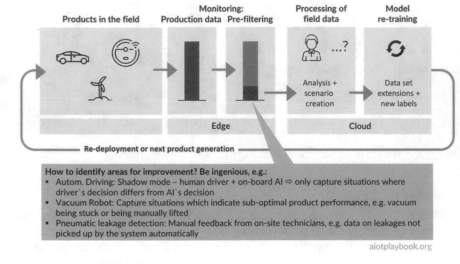

Fig. 21.9 The AIoT data loop

This means that in the AIoT Data Loop, data really is driving the development process. Marcus Schuster, project lead for embedded AI at Bosch, comments: *Data driven development will have the same impact on engineering as the assembly line had on production. Let's go about it with the necessary passion.*

21.1.8 Fifth: Productize the AI Approach

Based on the lessons learned from the Proof-of-Concept, the chose AI approach must now be productized so that it can support real-world deployment. This includes refining the model inputs/outputs, choosing a suitable AI method/algorithm, and aligning the AI model metrics with UX and IoT system requirements.

Model Inputs/Outputs

A key part of the system design is the definition of the model inputs and outputs. These should be defined as early as possible and without any ambiguity. For the inputs, it is important to identify early on which data are realistic to acquire. Especially in an AIoT solution, it might not be possible technically or from a cost point of view to access certain data that would be ideal from an analytics point of view. In the UBI example from above, the obvious choice would be to have access to the driving performance data via sensors embedded in the vehicle. This would either require that the insurance can gain access to existing vehicle data or that a new, UBI-specific appliance be integrated into the vehicle. This is obviously a huge cost factor, and the insurance might look for ways to cutting this, e.g., by requiring its customer to install a UBI app on their smartphones and try to approximate the driving performance from these data instead.

One can easily see that the choice of input data has a huge impact on the model design. In the UBI example, data coming directly from the vehicle will have a completely different quality than data coming from a smartphone, which might not always be in the car, etc. This means that UBI phone app data would require additional layers in the model to determine if the data are actually likely to be valid.

It is also important that all the information needed to determine the model output is observable in the input. For example, if very blurry photos are used for manual labeling, the human labeling agent would not be able to produce meaningful labels, and the model would not be able to learn from it [19].

Choosing the AI Algorithm

The choice of the AI method/algorithm will have a fundamental impact not only on the quality of the predictions but also on the requirements regarding data acquisition/data availability, data management, AI platforms, and skills and resources. If the AI method is truly at the core of the AIoT initiative, then these factors will have to be designed around the AI methods. However, this might not always be possible. For example, there might be existing restrictions with respect to available skills, or certain data management technologies that will have to be used.

Figure 21.10 provides an overview of typical applications of AI and the matching AI algorithms. The table is not complete, and the space is constantly evolving. When choosing an AI algorithm, it is important that the decision is not only based on the data science point of view but also simply from a feasibility point of view. An algorithm that provides perfect results but is not feasible (e.g., from the performance point of view) cannot be chosen.

Goals		Typical questions / examples	Algorithms
Determine category	Select between two categories	Is this A or B: "Does this image show a cat or not?"	Two-class classification algorithms, e.g.: ■ Support vector machine: Limited number of features ■ Averaged perceptron: fast training ■ Neural Networks: high accuracy
	Select between several categories	Is this A, B or C: "Is this a stop sign, a sign for incoming traffic or a do not enter sign?"	Multiclass classification algorithms, e.g.: ■ Logistic regression: short training times ■ Neural Network: high accuracy, long training times ■ Decision forest: high accuracy, short training times
Create recommendations			Recommender algorithms, e.g.: ■ Singular Value Decomposition (SVD): Collaborative filtering to capture patterns of interest
Predict values		„For how much money could I sell my house?"	Regression algorithms, e.g.: ■ Linear regression: short training times ■ Bayesian linear regression: small data sets ■ Decision forest regression: accurate, short training times ■ Neural network regression: accurate, long training times
Discover structures		„Can I group drivers based on their habits, e.g. acceleration, top speed or gender for different fees?"	Clustering algorithms, e.g.: ■ K-Means: simple to implement, but not good at identifying clusters of varying sizes and density
Detect anomalies		„Are there isolated extremly big values which could skew my model?"	Anomaly detection algorithms, e.g.: ■ One class SVM: small number of features ■ PCA: Short training times
Analyze media data	Text	Create automatic response to customer email	Text analytics algorithms, e.g.: ■ N-Gram, feature hashing, word2vector
	Images, video	Identify traffic signs on image	Image analytics algorithms, e.g.: ■ Densenet: high accuracy, efficient
	Sound	Identify potential leakages in pneumatic system	Audio analytics algorithms, e.g.: ■ Digitial Signal Processing, Filter Banks, Mel-Frequency Cepstral Coefficents

aiotplaybook.org

Fig. 21.10 AI selection matrix

In the context of an AIoT initiative, it should be noted that the processing of IoT-generated sensor data will require specific AI methods/algorithms. This is because sensor data will often be provided in the form of streaming data, typically including a time stamp that makes the data a time series. For this type of data, specific AI/ML methods need to be applied, including data stream clustering, pattern mining, anomaly detection, feature selection, multi-output learning, semi-supervised learning, and novel class detection [20].

Eric Schmidt, AI Expert at Bosch: *"We have to ensure that the reality in the field – for example the speed at which machine sensor data can be made accessible in a given factory – is matching the proposed algorithms. We have to match these hard constraints with a working algorithm but also the right infrastructure, e.g., edge vs. batch."*

Aligning AI Model Metrics with Requirements and Constraints

There are usually two key model metrics that have the highest impact on user experience and/or IoT system behaviour: model accuracy and prediction times.

Model accuracy has a strong impact on usability and other KPIs. For example, if the UBI model from the example above is too restrictive (i.e., rating drivers as more risk-taking than they actually are), than the insurance might lose customers simply because it is pricing itself out of the market. On the other hand, if the model is too lax, then the insurance might not make enough money to cover future insurance claims.

Eric Schmidt, AI Expert at Bosch: *"We currently see that there is an increasing demand in not only having accurate models, but also providing a quantification of the certainty of the model outcome. Such certainty measurements allow – for example – for setting thresholds for accepting or rejecting model results"*.

Similarly, in autonomous driving, if the autonomous vehicle cannot provide a sufficiently accurate analysis of its environment, then this will result (in the worst case) in an unacceptable rate of accidents, or (in the best case) in an unacceptable rate of requests for avoidable full brakes or manual override requests.

Prediction times tell us how long the model needs to actually make a prediction. In the case of the UBI example, this would probably not be critical, since this is likely executed as a monthly batch. In the case of the autonomous driving example, this is extremely critical: if a passing pedestrian is not recognized in (near-) real time, this can be deadly. Another example would be the recognition of a speed limited by an AIoT solution in a manually operated vehicle: if this information is displayed with a huge delay, the user will probably not accept the feature as useful.

21.1.9   Sixth: Release MVP

In the agile community, the MVP (Minimum Viable Product) plays an important role because it helps ensure that the team is delivering a product to the market as early as possible, allows valuable customer feedback and ensures that the product is viable. Modern cloud features and DevOps methods make it much easier to build on the MVP over time and enrich the product step-by-step, always based on real-world customer feedback.

For most AIoT projects, the launch of the MVP is a much "bigger deal" than in a pure software project. This is because any changes to the hardware setup – including sensors for generating data processed by an AI – are much harder to implement. In manufacturing, the term used is SOP (Start of Production). After SOP, changes to the hardware design usually require costly changes to the manufacturing setup. Even worse, changing hardware already deployed in the field requires a costly product recall. So being able to answer the question "What is the MVP of my smart coffee maker, vacuum robot, or electric vehicle" becomes essential.

Jan Bosch is Professor at Chalmers University and Director of the Software Center: *If we look at traditional development, I think the way in which you are representing the "When do I freeze what" is spot on. However, there is a caveat. In traditional development, I spend 90% of my energy and time obtaining the first version of the product. So I go from greenfield to first release, and I spend as little as possible afterwards. However, I am seeing many companies which are shifting toward a model that says "How do I get to a V1 of my product with the lowest effort possible?". Say I am spending 10% on the V1, then I can spend 90% on continuously improving the product based on real customer feedback. This is definitely a question of changing the mindset of manufacturing companies.*

Continuous improvement of software and AI models can be ensured today using a holistic DevOps approach, which covers all elements of AIoT: code and ML models, edge (via OTA) and cloud. This is discussed in more detail in the AIoT DevOps section.

Managing the evolution of hardware is a complex topic, which is addressed in detail in the Hardware.exe section.

Finally, the actual rollout or Go-to-Market perspective for AIoT-enabled solutions and products is not to be underestimated. This is addressed in the Rollout and Go-to-Market section.

21.1.10   Required Skills and Resources

AI projects require special skills, which must be made available with the required capacity at the required time, as in any other project situation. Therefore, it is important to understand the typical AI-roles and utilize them. Additionally, it is important to understand how the AI team should be structured and how it fits into the overall AIoT organization.

There are potentially three key roles required in the AI team: Data Scientist, ML Engineer, and Data Engineer. The Data Scientist creates deep, new Intellectual Property in a research-centric approach that can potentially require a 3 to 12-month development time or even longer. So the project will have to make a decision regarding how far a Data Science-centric approach is required and feasible, or in how far re-use of existing models would be sufficient. The ML Engineer turns models developed by data scientists into live production systems. They sit at the intersection of software engineering and data science to ensure that raw data from data pipelines are properly fed to the AI models for inference. They also write production-level code and ensure scalability and performance of the system. The Data Engineer creates and manages the data pipeline that is required for training data set creation, as well as feeding the required data to the trained models in the production systems (Fig. 21.11).

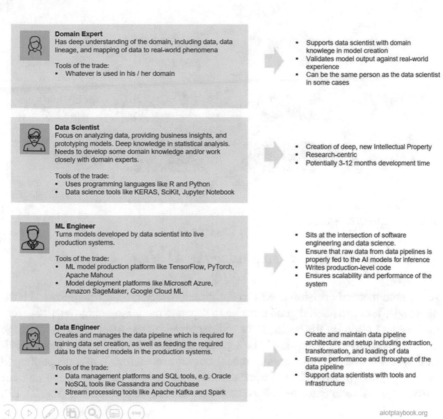

Fig. 21.11 AI roles for AIoT

Another important question is how the AI team works with the rest of the software organization. The Digital Playbook proposes the adoption of feature teams, which combine all the required skills to implement and deploy a specific feature. On the other hand, especially with a new technology such as AI, it is also important that experts with deep AI and data skills can work together in a team to exchange best practices. Project management has to carefully balance this out.

21.1.11   Model Design and Testing

In the case of the development of a completely new model utilizing data science, an iterative approach is typically applied. This will include many iterations of business understanding, data understanding, data preparation, modeling, evaluation/testing, and deployment. In the case of reusing existing models, the model tuning or – in the case of supervised learning models – data labeling should also not be underestimated (Fig. 21.12).

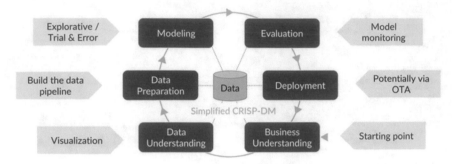

Fig. 21.12 Model development

21.1.12 Building and Integrating the AI Microservices

A key architectural decision is how to design microservices for inference and business logic. It is considered good practice to separate the inferencing functions from the business logic (in the backend, or – if deployed on the asset – also in the edge tier). This means that there should be separate microservices for model input provisioning, AI-based inferencing, and model output processing. While decoupling is generally good practice in software architecture, it is even more important for AI-based services in case specialized hardware is used for inferencing (Fig. 21.13).

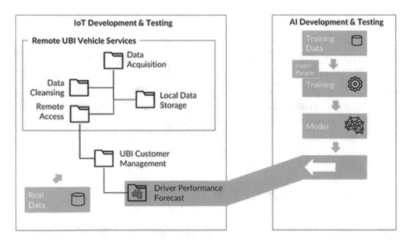

Fig. 21.13 UBI microservices

21.1.13   Setting Up MLOps

Automating the AI model development process is a key prerequisite not only from an efficiency point of view, but also for ensuring that model development is based on a reproducible approach. Consequently, a new type of DevOps is emerging: MLOps. With the IoT, MLOps not only have to support cloud-based environments but also potentially the deployment and management of AI models on hundreds – if not hundreds of thousands – of remote assets. In the *Digital Playbook* there is a dedicated section on Holistic DevOps for AIoT because this topic is seen as so important (Fig. 21.14).

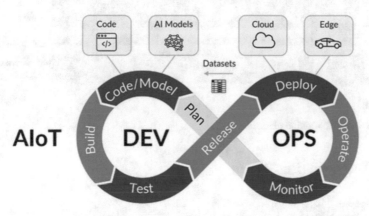

Fig. 21.14 Holistic DevOps for AIoT

21.1.14   Managing the AIoT Long Tail: AI Collaboration Platforms

When addressing the long tail of AI-enabled opportunities, it is important to provide a means to rapidly create, test and deploy new solutions. Efficiency and team collaboration are important, as is reuse. This is why a new category of AI collaboration platforms has emerged, which addresses this space. While high-end products on the short tail usually require very individual solutions, the idea here is to standardize a set of tools and processes that can be applied to as many AI-related problems as possible within a larger organization. A shared repository must support the workflow from data management over machine learning to model deployment. Specialized user interfaces must be provided for data engineers, data scientists and ML engineers. Finally, it is also important that the platforms support collaboration between the aforementioned AI specialists and domain experts, who usually know much less about AI and data science (Fig. 21.15).

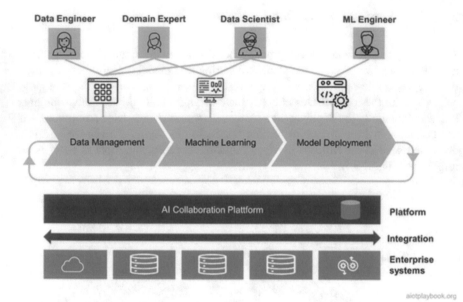

Fig. 21.15 AI collaboration platform

21.2   Data.exe (Fig. 21.16)

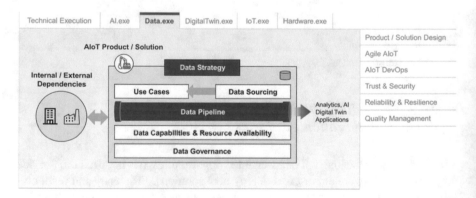

Fig. 21.16 Overview of Ignite AIoT framework

As part of their digital transformation initiatives, many companies are putting data strategies at the center stage. Most enterprise data strategies are a mixture of high-level vision, strategic principles, goal definitions, priority setting, data governance models, architecture tools and best practices for managing semantics and deriving information from raw data.

Since both AI and IoT are also very much about data, every AIoT initiative should also adopt a data strategy. However, it is important to note that this data strategy must work on the level of an individual AIoT-enabled product or solution, not the entire enterprise (unless, of course, the enterprise is pretty much built around said product/solution). This section of the AIoT Framework proposes a structure for an AIoT Data Strategy and identifies the typical dependencies that must be managed.

21.2.1   Overview

The AIoT Data Strategy proposed by the AIoT Framework is designed to work well for AIoT product/solution initiatives in the context of a larger enterprise. Consequently, it focuses on supporting product/solution implementation and long-term evolution and tries to avoid replicating typical elements of an enterprise data strategy (Fig. 21.17).

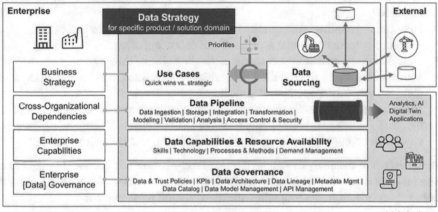

Fig. 21.17 AIoT data strategy

The AIoT Data Strategy has four main elements. First, the development of a prioritization framework that aims to make the relationship between use cases and their data needs visible. Second, management of the data-specific implementation aspects, as well as the Data Lifecycle Management. Third, Data Capabilities required to support the data strategy. Fourth, a lean and efficient Data Governance approach was designed to work on the product/solution level.

Of course, each of these four elements of the AIoT Data Strategy has to be seen in the context of the enterprise that is hosting product/solution development: Enterprise Business Strategy must be well aligned with the use cases. Data-specific implementation projects frequently have to take cross-organization dependencies into consideration, e.g., if data are imported or exported across the boundaries of the current AIoT product/solution. Product/solution-specific data capabilities must be aligned with the existing enterprise capabilities. Product/solution-specific data governance always has to take existing enterprise-level governance into consideration.

21.2.2 Business Alignment & Prioritization

The starting point for business alignment and prioritization should be the actual use cases, which are defined and prioritized by business sponsors, or Epics which have been prioritized in the agile backlog. Sometimes, Epics might be too coarse grained. In this case, Features can be used instead.

For each Use Case/Epic, an analysis from the data perspective should be completed:

- What are the actual data needs to support the Use Case/Epic?
- Which of these data is believed to be already available, which must be newly acquired?
- How can the required data quality be ensured for the particular use case?
- What are potential financial aspects of the data acquisition?
- How do the use cases support the monetization side of things?
- Is this a case where the required data adds functional value to the use case, or is there a direct data monetization aspect to it?
- What are the relationships between the identified data and the other elements of the AIoT Data Strategy: Implementation & Data Lifecycle Management, specific capabilities applying to this particular kind of data, and Data Governance?

A key aspect of the analysis will be the **Data Acquisition** perspective. For data that can (at least theoretically) be acquired within the boundaries of the AIoT product/ solution organization, the following questions should be answered:

- Is the required technical infrastructure already available?
- Does the team have the required capabilities and resources available?
- Especially in the case of AIoT data acquired via sensors:

  - Are new sensors required?
  - If so, what is the additional development and unit cost?
  - Is there an additional downstream cost from the asset/sensor line-fit point of view (i.e. additional manufacturing costs)?
  - What is the impact on the business plan?
  - What is the impact on the project plan?
  - What are the technical risks for new, unknown sensor technologies?
  - What are required steps in terms of sourcing and procurement?

For data that need to be acquired from other business units, a number of additional questions will have to be answered:

- Is it technically feasible to access the data (availability of APIs, bandwidth, support of required data access frequency and volume, etc.)?
- Can the neighboring business unit support your requirements, not only in terms of technical access, but also in terms of project support and timelines?
- Are there costs involved in technical implementation and/or data access (internal billing)?
- Are there potential limitations or restrictions due to existing internal data governance guidelines, regional or organizational boundaries, etc.?

For data that have to be acquired from external partners or suppliers, there are typically a number of additional complexities that will have to be addressed:

- Technical feasibility across enterprise boundaries
- Legal framework required for data access
- SLA insurance
- Billing and cost management

Based on all of the above, the team should be able to assess the overall feasibility and costs/efforts involved on a per use case/per data item basis. This information is then used as part of the overall prioritization process.

21.2.3 Data Pipeline: Implementation & Data Lifecycle Management

Sometimes it can be difficult to separate data-specific implementation aspects from general implementation aspects. This is an issue that the AIoT Data Strategy needs to deal with to avoid redundant efforts. Typical data-specific implementation and Data Lifecycle Management aspects include the following:

- Data Ingestion: In our context, data ingestion should first be seen as moving data from outside of our organization's boundary to within. Second, technical aspects such as stream vs. batch processing need to be addressed. Typical data ingestion tasks also include cleansing and quality assurance.
- Storage: Depending on the business and technical requirements, data can be stored permanently or temporarily, structured or unstructured, with or without backup, with cache-only or with operational/transactional support, etc. This often needs to be addressed differently for different data types.
- Integration: Data integration is the process of merging data from different sources into a single, unified view. In the case of AIoT, this can be – for example – sensor data fusion, done close to the sensors in the edge layer. Or it can be – usually on a high-level of abstraction – a real-time data stream integration process. Or it can be – typically further in the backend – a batch-oriented integration process.
- Transformation: Many projects spend much time with data transformation, since this is often a prerequisite for data integration or further data processing. The

approaches chosen usually vary widely depending on the format, structure, complexity, and volume of the data being transformed.

- Modeling: Data modeling is usually a key step toward dealing with semantics of data and deriving information from raw data. There are different levels of data modeling, including conceptual, logical and physical levels. Another important type of model building on top of data models is AI/ML models. However, these models are usually less data-structure oriented and more mathematical/statistical models.
- Validation: Data validation is the tool that helps ensure data quality, e.g., by applying data cleansing and validation checks. Data validation can use simple, local "validation rules" or "validation constraints" that check for correctness and meaningfulness (e.g., a date of birth cannot be in the future). In some cases, data validation can actually be much more complex, e.g., involving interactions with remote systems, or even AI/ML-based validation algorithms.
- Analysis: In many cases, data analysis is a key use case other than, for example, transactional use of the data. Generally, data analysis supports the discovery of useful information and supports decision-making. Data analysis is a multifaceted topic. It is key that the required Data Capabilities are provided to support here.
- Access Control & Security: Finally, effectively ensuring confidentiality and secure handling of data must be part of every AIoT data strategy. This includes both IoT data coming from assets and data combining from users, other business units, or event external data sources. While security is sometimes dealt with on a different level, fine-grained data access control must usually be dealt with as part of the data strategy.

Another key aspect of Implementation & Data Lifecycle Management is dealing with cross-organizational dependencies. While the earlier data acquisition phase might have already answered some of the high-level questions related to this topic, on the implementation level efficient stakeholder management is a key success factor. Often, earlier agreements with respect to technical data access or commercial conditions, will have to be reviewed, revised or refined during the implementation phase. Some practitioners say that this can sometimes be more difficult in the case of cross-divisional data integration within one enterprise than across enterprise boundaries.

21.2.4 Data Capabilities and Resource Availability

Data-related capabilities can be important in a number of different areas, including:

- Skills: Data-related skills can include a number of areas, including specific data-processing technologies and mathematical, statistical, or algorithmic skills in AI/ML, etc.
- Technology: For an AIoT product/solution initiative, it is usually important that technical management agrees on fixed setup technologies that cover most of the required use cases, e.g., batch vs real-time processing, basic analytics vs AI/ML, etc.

-

- Processes & Methods: Depending on the specific environment, this can also be a very important aspect. Data-related processes and methods can be specific to a certain analytics method, or they can be related to certain processes and methods defined by an enterprise organization as mandatory.

Depending on the project requirements, it is also important that specific capabilities be supported by appropriate resources. For example, if it is clear that an AIoT project will require the development of certain AI/ML algorithms, then the project management will have to ensure that this particular capability is supported by skilled resources that are available during the required time period. Managing the availability of such highly specialized resources is a topic that can be difficult to align with the pure agile project management paradigm and might require longer-term planning, involving alignment with HR or sourcing/procurement.

21.2.5 Data Governance

Larger AIoT product/solution initiatives will require Data Governance as part of their Data Strategy. This Data Governance cannot be compared with a Data Governance approach typically found on the enterprise level. It needs to be lightweight and pragmatic, covering basic aspects such as:

- Data & Trust Policies: How is this specific AIoT product/solution dealing with this topic? This is likely to be very use case specific, so the AIoT initiative will have to build on generic enterprise-level requirements but will have to add policies specific to its own use case.
- Data Architecture: It is not always clear if data architecture is a discipline on its own, or if this is simply one facet of the product/solution architecture. For example, the AIoT Framework has a dedicated viewpoint to support the combination of data and functionality.
- Data Lineage: Data lineages traces where data originate, what happens with it on the way, and where it moves over time. Data lineage provides visibility and transparency and can help simplify root cause analysis in the data analytics process. Data Governance can either support the central documentation of data lineages or provide tools and best practices for implementation teams.
- Metadata Management and Data Catalog: Efficient management of metadata is a prerequisite for efficient data processing and analytics. Types of metadata include descriptive, structural and administrative. A data catalog can provide support for metadata management, together with other tools, such as search.
- Data Model Management: For many AIoT applications, centrally managing a high-level data model that describes key entities and their relationships, as well as dependencies on different use cases and components, can be of great help in creating transparency and improving alignment between different teams. The AIoT Framework proposes a lightweight AIoT Domain Model approach. In addition, the Data Governance team could also provide tooling and best practices

for teams that need more detailed models in their areas. This can also be linked back to the Metadata Management and Data Catalog topics.

- API Management: In his famous "API Mandate", Amazon CEO Jeff Bezos declared that "*All teams will henceforth expose their data and functionality through service interfaces.*" at Amazon. This executive-level support for an API-centric way of dealing with data exchange (and exposing component functionality) shows how important API management has become at the enterprise level. The success of an AIoT initiative will also depend strongly on it. If there is no enterprise-wide API infrastructure and management approach available, this is a key support element that must be provided and enforced by the Data Governance team.

Finally, the Data Governance/Data Strategy team should give itself a setup of KPIs by which they can measure their own success and the effectiveness and efficiency of the AIoT Data Strategy.

21.3   Digital Twin.exe (Fig. 21.18)

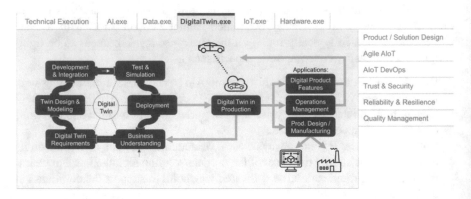

Fig. 21.18   Digital Twin

As discussed in Digital Twin 101, a Digital Twin is the virtual representation of a real-world physical object. Digital Twins help manage complexity by providing a semantically rich abstraction layer, especially for systems with a high level of functional complexity and heterogeneity. As an AIoT project lead, one should start by looking at the question "Is a Digital Twin needed, and if so – what kind of Digital Twin?" before defining the Digital Twin implementation roadmap.

21.3.1   Is a Digital Twin Needed?

The decision of whether and when to apply the Digital Twin concept in an AIoT initiative will depend on at least two key factors: Sensor Data Complexity/Analytics Requirements and System Complexity (e.g., the number of different machine types, organizational complexity, etc.).

If both are low, the system will probably be fine with using Digital Twin more as a logical design concept and applying traditional data analytics. Only with increasing sensor data complexity and analytics requirements will the use of AI be required.

High system complexity is an indicator that dedicated Digital Twin implementation should be considered, potentially utilizing a dedicated DT platform. The reason is that a high system complexity will make it much harder to focus on the semantics. Here, a formalized DT can help (Fig. 21.19).

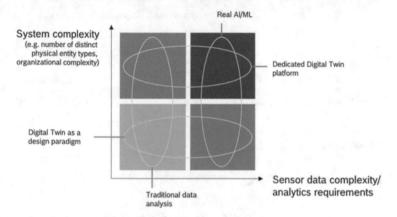

Fig. 21.19 Conclusions – digital Twin and AIoT

21.3.2 If So, What Kind of Digital Twin?

Since Digital Twin is a relatively generic concept, the concrete implementation will heavily depend on the type of data that will be used as the foundation. Since Digital Twins usually refer to physical assets (at least in the context of our discussions), the potential data can be identified along the lifecycle of a typical physical asset: design data or digital master data, simulation data, manufacturing/production data, customer data, and operational data. For the operational data, it is important to differentiate between data related to the physical asset itself (e.g., state, events, configuration data, and history) versus data relating to the environment of the asset.

Depending on the application area, the Digital Twin (DT) can have a different focus. The Operational DT will mainly focus on operational data, including the internal state and data relating to the environment. PLM-focused DT will combine the product/asset design perspective with the operational perspective, sometimes also adding manufacturing-related data. The simulation-focused DT will combine design data with operational data and apply simulation to it. And finally, the holistic DT will combine all of the above (Fig. 21.20).

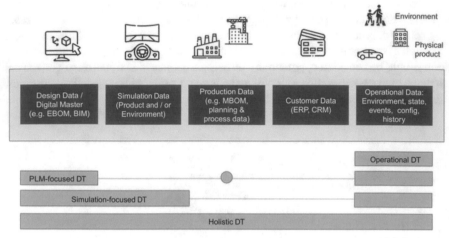

Fig. 21.20 DT categories

21.3.3 Examples

The Digital Twin concept is quite versatile, and can be applied to many different use cases. Figure 21.21 provides an overview of four concrete examples and how they are mapped to the DT categories introduced earlier.

Type	Design data	Simulation	Production data	Customer data	Operational data	Example
Operational Digital Twin				✓	✓	**Pneumatic system: leakage detection** Digital Twin used predominantly to model customer site and key pneumatic components on site, including documentation of potential leaks
PLM-focused Digital Twin	✓				✓	**Drone-based building façade inspection** This system by TÜV SÜD is using drones to scan building facades. The scans are checked for potential problems using AI. The resulting problem report is merged back with the BIM-based design model of the building.
Simulation-focused Digital Twin	✓	✓			✓	**Physics simulation for high-power inverter** Case study from Bosch Bosch Engineering Battery Management Systems (see Digital Twin 101), using physics simulation and virtual sensors to improve accuracy of system monitoring and predictive maintenance algorithms.
Holistic Digital Twin	✓	✓	✓	✓	✓	**Digital Twin for elevator lifecycle management** This example shows the use of a digital twin across the entire lifecycle of an elevators, including product design, custom fitting, monitoring, maintenance, and continuous product optimization

aiotplaybook.org

Fig. 21.21 Examples for different DT categories

The drone-based building facade inspection is covered in detail in the TÜV SÜD case study. The physics simulation example is covered in the Digital Twin 101 section. The following provides an overview of the pneumatic system example, as well as the elevator example.

Operational DT: Pneumatic System

Leakage detection for pneumatic systems is a good example for an operational Digital Twin. Pneumatic systems provide pressured air to implement different use cases, e.g., the drying of cars in a car wash, eliminating bad grains in a stream of grains analyzed using high-speed video data analytics, or cleaning bottles in a bottling plant. Experts estimate that pneumatic systems consume 16 billion kilowatt hours annually, with a savings potential of up to 50% (mader.eu). In order to address this savings potential, an AIoT-enabled leakage detection system can help to identify and fix leakages at customer sites. One such solution is currently developed by the AIoT Lab. This solution is based on a combination of ultrasound sensors and edge-ML for sound pattern analysis. The solution can be used on-site to perform an analysis of the customer's pneumatic application for leakages. The results can then be used by a service technician to fix the problems and eliminate the leakages (Fig. 21.22).

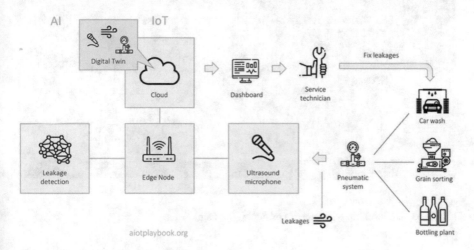

Fig. 21.22 Example: digital Twin for pneumatic system

The foundation for the leakage detection system is an operational Digital Twin. Since customers usually don not provide detailed design information about their own systems, the focus here is to obtain as much information during the site visit and build up the main part of the Digital Twin dynamically while being on site. The system is based on Digital Twin data in four domains:

- Domain I includes the components of the AIoT solution itself, e.g., the mobile gateways and ultrasound sensors. This DT domain is important to support the system administration, e.g., OTA-based updates of the ML models for sound detection.

- Domain II includes the pneumatic components found on-site, including pressure generators, pressure tanks, valves, etc. The definitions of these components are provided via the product catalogue, and can be selected dynamically on-site.
- Domain III includes the fuselage and how it is mapped to the applications of the customer. Key parts of the customer equipment must be identified and included in the DT model for documentation purposes. Usually, only those parts of the customer equipment are captured that are involved with any of the leakages found.
- Domain IV includes the leakages that are identified during the on-site assessment. These leakages are also captured as Digital Twins, including information about the related sound patterns, as well as the position of the leakage relative to DT information from domains II and III.

The creation of the Digital Twins happens along these domains: DT data in domain I are created once per test equipment pack. Domains II-IV are created dynamically and per customer site (Fig. 21.23).

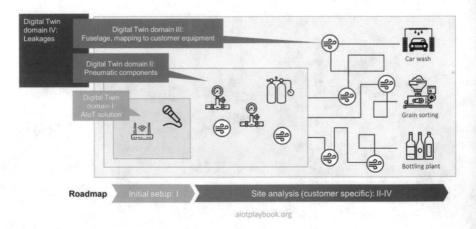

Fig. 21.23 Digital Twin – pneumatics example – domains

Daniel Burkhardt, Chief Product Owner, AIoT Lab: *We have the goal of providing a solution architecture that enables ML model reuse and holistic AIoT DevOps. The implementation of leakage detection based on a Digital Twin of a pneumatic system provided us with relevant insights about the requirements and design principles for achieving this goal. In comparison to typical software development, reuse and AIoT DevOps require design principles such as continuous learning, transferability, modularization, and openness. Realizing these principles will guarantee the ease of use of AIoT for organizations with, e.g., no technological expertise, which in the long term leads to more detailed and meaningful Digital Twins and thus more accurate and valuable analytics.*

Holistic Digital Twin: DT and Elevators
A good example of the use of a holistic Digital Twin approach is elevators, since they have a quite long and complex lifecycle that can benefit from this approach. What is interesting here as well is the combination of the elevator lifecycle in combination with the building lifecycle, since most elevators are deployed in buildings. The example in the following shows how a standard elevator design is fitted into a building design. This is a complex process that needs to take into consideration the elevator design specification, building design, elevator shaft design, and required performance parameters (Fig. 21.24).

Fig. 21.24   Digital Twin of building and elevator – 3D model

The CAD model and EBOM data of the elevator design can be a good foundation for the digital twin. To support efficient monitoring of the elevator during the operations phase, an increasing number of advanced sensors have been applied. These include, for example, sensors to monitor elevator speed, braking behavior, positioning of the elevator in the elevator shaft, vibrations, ride comfort, doors, etc. Based on these data, a dashboard can be provided that provides reports for the physical conditions and the elevator utilization.

One pain point for building operators is the usually mandatory on-site inspections by a third party inspection service. Using advanced remote monitoring services based on a digital twin of the elevator, some countries are already allowing combination or remote and on-site inspections. For example, instead of 12 on-site inspections per year, this could be reduced to 4 on-site inspections with 8 inspections being performed remotely. This helps save costs and reduces operations interruptions due to inspection work.

The Digital Twin concept helps brings together all relevant data, and allows semantic mappings between data from different perspectives and created during different stages of the lifecycle (Fig. 21.25).

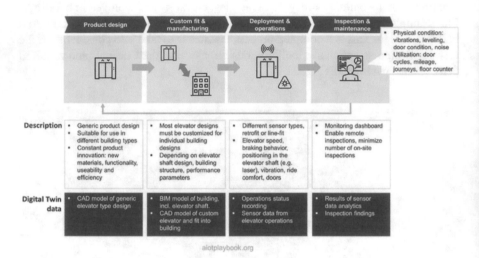

Fig. 21.25 Holistic DT example – elevator

21.3.4 Digital Twin Roadmap

From the execution perspective, a key question is how to design a realistic roadmap for the different types of Digital Twins we have looked at here. The following provides two examples, one from the automotive perspective and one from the building perspective.

Operational Digital Twin (Vehicle Example)
Let us assume an OEM wants to introduce the Digital Twin concept as part of their Software-defined Vehicle initiative. Over time, all key elements of the vehicle should be represented on the software layer as Digital Twin components. How should this be approached?

Importantly, this should be done step by step or more precisely use case by use case. Developing a Digital Twin for a complex physical product can be a huge effort. The risk of doing this without specific use cases and interim releases is that the duration and cost involved will lead to a cancellation of the effort before it can be finished. This is why it is better to select specific use cases, then develop the Digital Twin elements required for them, release this, and show value creation along the way. Over time, the Digital Twin can then develop to an abstraction layer that will cover the entire asset, hopefully enabling reuse for many different applications and use cases (Fig. 21.26).

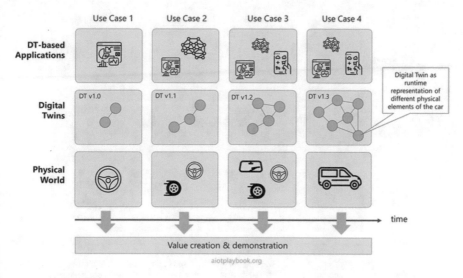

Fig. 21.26 DT evolution

Holistic Digital Twin (Building Example)

A good example of use of a holistic Digital Twin concept from design to operation and maintenance is the digital building lifecycle:

- During the building design phase, the BIM (Building Information Model) approach can help optimize the design with simulation and automated validation. This way, aspects such as future operational sustainability and capacity can be evaluated. Automated design validation provides a higher level of planning safety.
- During the building construction process, AIoT-enabled solutions such as robot-based construction progress monitoring can provide transparency and reliability. Meeting budgets and timelines can be better ensured.
- Sub-systems like elevators can also be integrated into the Digital Twin approach, as discussed in the previous section.
- Finally, building inspection can be supported by solutions such as the Drone-based façade inspection. The results of the façade inspection can be mapped back to the Digital Twin, augmenting the planning data with real-world as-is data.

The decision for a BIM/Digital Twin-based approach for building and construction is strategic. Upfront investments will have to be made, which must be recuperated through efficiency increased further downstream. The holistic Digital Twin approach here is promising, but requires a certain level of stringency to be successful (Fig. 21.27).

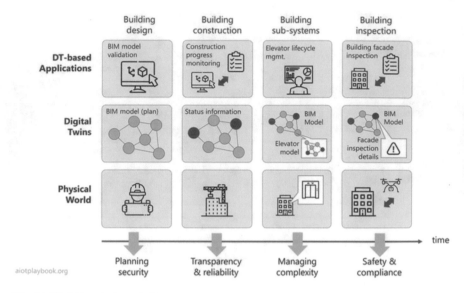

Fig. 21.27 DT evolution – building example

21.3.5   Expert Opinion

The following short interview with Dominic Kurtaz (Managing Director for Dassault Systèmes in Central Europe) highlights the experience that a global PLM company is currently making with its customers in the area of AIoT and Digital Twins.

Dirk Slama: *Welcome Dominic. Can you briefly introduce your company?*

Dominic Kurtaz: *Dassault Systèmes consists of 20,000 inspired people around the world, developing software solutions and supporting clients in the manufacturing, healthcare and life science sector, as well as the infrastructure sector. We help to digitally design and manufacture more than 1 in 4 of the physical products you touch every day, with a focus on how they are being used by the end users and consumers. We believe that the virtual world can enhance and improve the overall physical world toward a more sustainable world, which I think is probably a good segue to the whole topic of AIoT.*

Dirk: *In this context, AIoT and Digital Twins can play an important role as enablers. What kind of activities and investments are you currently seeing in this space?*

Kurtaz: *When people think of AIoT or IoT, they immediately think of operational performance measurements with sensors, predictive maintenance, and so on. Which of course is a very valid application, but we need to think far beyond that. This is why I like this concept of a holistic Digital Twin. We need to take a step back from IoT right now. When you are looking at the Experience Economy, you will see that the value that we perceive as customers and consumers is going increasingly away from the actual product itself. Today, it is often much more* about the end-to-end experience: how the product is perceived how we select it, how we are using it, and how we dispose and recycle the product. The end-to-end life cycle experience is clearly important. From my experience, we need to look at the IoT through the eyes of the customer and the eyes of the consumer. First, we have to understand how business strategies and business execution with AIoT can truly support and improve those aspects.

Second, I believe that the Digital Twin is truly becoming pervasive across industries and all products. Take, for example, one of the most mundane products that we experienced in our lives – the light bulb. If you go back 10 years, it was just this item at the end of the shopping list that you grab off from the shop shelf without thinking much about it – you bought it, you screwed it in, you turned it on and off, and hopefully you would never have to think about that product again for the next years…until it breaks.

Today, this is fundamentally changing. I am not just buying a commodity product for my house anymore. I am buying something that is part of a connected ecosystem. I can set different moods at home using different light configurations. I can use smart lighting as part of my home security system. From a business perspective, this is a game changer. Light bulb manufacturers are no longer just producing light bulbs – today, they are connected to their customers. In the past, we did not know our customers or how they were using our products. Today, – enabled by the IoT – I can have a direct relationship with my customer. This will change things on many levels and opens up new business models.

Thus far, we have only seen the tip of the iceberg: although many of the enabling technologies are reaching a good level of maturity, the actual implementations are often still very immature and limited to those basic connectivity features – but not delivering the holistic Digital Twin experience. For example, I have recently bought a new kitchen, including connectivity to my smart home. Now I can control and integrate it into my own kitchen facilities. This is really good and interesting as well as delivering additional features but I was not able to experience and understand the value that it can really bring until after I had purchased all of those IoT enabled and connected products. And in today's world, I should have been able to use a Digital Twin of the product prior to my buy to fully understand not just the product, but the behavior, the context, the operational aspect of that post my buying – and that is simply not yet possible. Take, as another example, mobility. As a customer, I should be able to experience all these new features such as advanced driver assistance, before I acquire the physical product – enabled by a holistic Digital Twin. I really want to be able to experience in the virtual world how these products are going to behave, before using them in the physical world. This is also very helpful for product development, because it allows us to validate the customer experience in the virtual world – before making expensive investments in physical prototypes.

From what I am seeing from our customers, this is not just a hype or a fad. I think it is absolutely mission critical for anybody who is designing and manufacturing products, and dealing with the digital experience of those products. We see this *across all industries where we are operating: manufacturing, healthcare, life science, and infrastructure.*

Dirk Slama: *What are your recommendations from the implementation perspective?*

Dominic Kurtaz: *You need a clear focus on the end user experience that you are trying to deliver. This will determine the holistic design philosophy you need to apply. Many companies have started with Big Data, and they are now drowning in it. The problem is to find and connect the data that are relevant for the end user experience. The connection of digital, semantic models with data will open up potential for all industries. Of course, this has to be done step-by-step, use case by use case – building up the holistic Digital Twin with a clearly value-driven approach.*

Another key aspect is the alignment between the digital supply chain and the physical supply chain. For the IT, we have Continuous Integration and Continuous Delivery (CI/CD). For the physical product, we have simultaneous engineering and closed loop PLM. The challenge is now to close the even bigger loop around all of this– bringing IT DevOps together with physical product engineering. This is exactly where AIoT and Digital Twin will play an important role. AIoT enables new digital/physical product features. And the Digital Twin is the semantic interface between the digital and the physical world. During design and development, the Digital Twin helps create the required interfaces at the technical and the organizational levels. During runtime, it enables a new customer experience.

21.4   IoT.exe (Fig. 21.28)

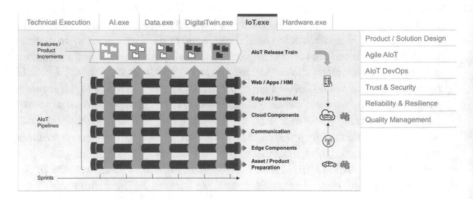

Fig. 21.28 Ignite AIoT – Internet of Things perspective

The IoT perspective in AIoT is usually much more focused on the physical product and/or the site of deployment, as well as the end-to-end system functionality. In this context, it makes sense to look at the IoT through the lens of the process that will support building, maintaining and enhancing the end-to-end system functionality.

The AIoT Framework is based on the premise that overall an agile approach is desirable, but that due to the specifics of an AIoT system, some compromises will have to be made. For example, this could concern the development of embedded and hardware components, as well as safety and security concerns.

Consequently, the assumption is that there is an overarching agile release train, with different (more or less) agile work streams. Each workstream represents some of the key elements of the AIoT system, including cloud services, communication services and IoT/EDGE components. In addition, AIoT DevOps & Infrastructure as well as cross-cutting tasks such as security and end-to-end testing are defined as workstreams. Finally, asset preparation is a workstream that represents the interface to the actual asset/physical product/site of deployment.

The following provides a more detailed description of each of the standard work streams:

- Agile Release Train: Responsible for end-to-end coordination, UX, and system architecture; ultimately responsible for ensuring that the AIoT system is implemented, tested, deployed and released
- Cross-Cutting: Addresses tasks that are cutting especially across the cloud and IoT/EDGE, including end-to-end security, testing and QA
- AIoT DevOps & Infrastructure: Must provide the infrastructure and processes for automating the AIoT system lifecycle, utilizing the AIoT DevOps concepts outlined in the AIoT Framework
- Cloud Services: Should more accurately be called *Backend services, including cloud and on-premises AIoT applications, as well as enterprise system integration/EAI*. Must also address the backend side of Digital Twin, as well as AIoT-related business processes
- Communication Services: Must provide LAN and WAN communication services. Can involve complex service contract negotiations in case a global AIoT WAN is required
- IoT/EDGE Components: Includes responsibility for the development/procurement of all hardware (e.g., gateways, sensors), software, firmware and AI/ML execution environments deployed on or near the asset/product
- Asset Preparation: Must ensure that the asset/physical product (or, in the case of an AIoT solution, the sites of deployment) are prepared to work with the AIoT system. Must include basic tasks such as ensuring power supply and providing storage/assembly points for AIoT hardware components

The following will look at both the product and solution perspectives in more detail.

21.4.1   Digital OEM: Product Perspective

This section looks at key milestones for an AIoT-enabled product, along the work streams defined earlier:

- Basic prototype/pilot: Must include a combination of what will later become the AI-enabled functionality (could be scripted/hard coded at this stage), plus basic system functionality and ideally a rudimentary prototype of the actual asset/ physical product (*A/B samples*). Should show end-to-end how the different components will interact to deliver the desired user experience

- Fully functional prototype: Functional, basic prototype with full AIoT functionality and a relatively high level of maturity. Must include first real AI models and AI-driven functionality, as well as full asset/physical product functionality (*C/D samples*). After this, both the APIs between the cloud and EDGE should be stable, as well as the interfaces to the asset/physical product (power lines, antenna and gateway fastening, etc.).
- AIoT MVP: This focuses only on the AIoT elements, assuming that the asset/ physical product will no longer undergo any major changes. The AIoT MVP must not only be functionally complete, but also ensure that all procurement aspects are finalized. Furthermore, a fully automated AIoT DevOps infrastructure, including cloud, IoT and AI pipelines, should be developed
- SOP (Start of Production): This is the day of no return: the manufacturing lines will now start processing assets/physical products and shipping them to customers around the world. Any changes/fixes on the hardware side will now become very costly or nearly impossible. Currently, the required operations support must also be fully operational (either providing fully automated online support services, or call-center or even on-site field services)
- Cloud SW Updates after SOP: This must utilize the AIoT DevOps pipeline, including Continuous Integration and Continuous Testing for quality purposes
- EDTE SW Updates after SOP: Finally, this must utilize the established OTA infrastructure to deliver updates to assets in the field (which will have already been established in the later stages of system field tests)

Note that this perspective does not differentiate between the hardware engineering and manufacturing perspectives of the on-asset AIoT hardware vs. the actual asset/ physical product itself. Furthermore, it also does not differentiate between line-fit and retrofit scenarios (Fig. 21.29).

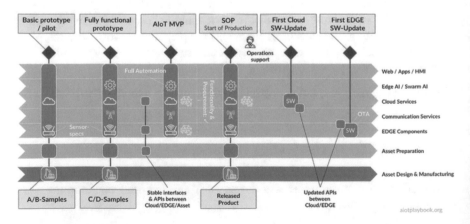

Fig. 21.29 AIoT product perspective

21.4.2   Digital Equipment Operator: Solution Perspective

An AIoT solution is usually not focused on the design/manufacturing of assets/
physical products. In many cases, assets are highly heterogeneous, and the AIoT
solution components will be applied using a retrofit approach. Instead of asset preparation, the focus is on site preparation. Additionally, the level of productization is
usually not as high.

This makes the process and the milestones easier and less complex:

- Pilot: Usually, much more lightweight; could simply be some sensors retrofitted
  to an existing asset, with a WLAN connection to a standard cloud backend
- MVP: Again, more lightweight and most likely also less sophisticated in terms
  of process automation
- Roll-out: Critical part of the process: not only in technical terms but also in terms
  of fulfilling on-site user expectations
- First Cloud SW-Update: Should be automated, utilizing existing standard cloud
  DevOps mechanisms
- First EDGE SW-Update: Can be automated and utilizing OTA, but for small-
  scale solutions; potentially also manual (Fig. 21.30)

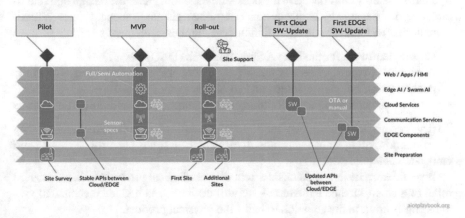

Fig. 21.30 AIoT solution perspective

21.5   Hardware.exe (Fig. 21.31)

The execution of the hardware implementation can vary widely. For a simple retrofit
solution using commercial-off-the-shelf hardware components, this will mainly be
a procurement exercise. For an advanced product with complex, custom hardware,
this will be a multidisciplinary exercise combining mechanical engineering, electric
and electronic engineering, control system design, and manufacturing.

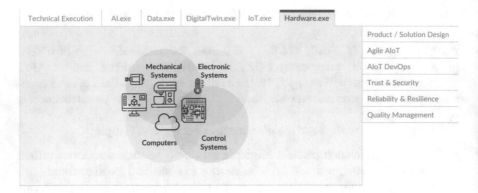

Fig. 21.31 AIoT hardware

21.5.1 A Multidisciplinary Perspective

The development of custom hardware often requires a multidisciplinary perspective. Take, for example, the development of the predictive maintenance solution for hydraulic systems, introduced in the case study section. Here, the design and manufacturing of the actual hydraulic system components is not in scope. However, hardware design and manufacturing still include a number of elements:

- Custom hardware for the Data Acquisition Hub (DAQ)
- A number of custom sensor packages to monitor electric motors, hydraulic pumps, tanks, oil quality, filters, and so on
- Custom connecting elements for fitting the sensors onto the hydraulic components

To develop this hardware, a number of different skills are required, including strong domain knowledge, knowledge about electronic systems, control systems, and embedded compute nodes.

If we go even further and consider real digital/physical products – like a vacuum robot or a smart kitchen appliance – we will even need to include mechanical systems engineering in the equation to build the physical product.

Mechatronics is the discipline that brings all these perspectives together, combining mechanical system engineering, electronic system engineering, control system engineering and embedded as well as general IT system engineering. The intersection between mechanical systems and electronic systems is often referred to as electromechanics. The intersection between electronic systems and control systems includes control electronics. The intersection between control systems and computers includes digital control systems. Mechanical systems usually require mechanical CAD/CAM for system design and modelling, as well as validation via simulation. Model Based System Engineering (MBSE) supports this with collaboration platforms covering system requirements, design, analysis, verification and validation (Fig. 21.32).

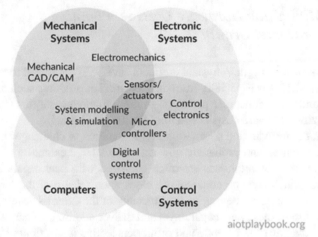

Fig. 21.32 Mechatronics – a multidisciplinary perspective

21.5.2 *Embedded Hardware Design and Manufacturing*

Embedded hardware design and manufacturing are often at the heart of AIoT development because even for a retrofit solution, this is often a key requirement. Even if standard microprocessors, CPUs, sensors and communications modules are used, they often have to be combined into a custom design to exactly fit the project requirements. During the planning phase, hardware requirements are captured in a specification document. The analysis and design phase includes feasibility assessment, schematic PCB (Printed Circuit Board) design and layout, and BOM (Bill of Material) optimization. Procurement should not be underestimated, including component procurement and supply chain setup. The actual board bring-up includes hardware assembly, software integration, testing and validation, and certification. Manufacturing preparation includes machine configuration, assembly preparation, as well as automated inspection. After the SOP (Start of Production), logistics and shipment operations as well as customer support will have to be ensured (Fig. 21.33).

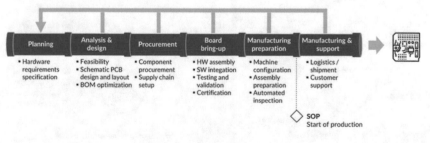

Fig. 21.33 Embedded hardware design and manufacturing

21.5.3   Minimizing Hardware Costs vs. Planning for Digital Growth

In the past, almost all digital/physical products have been optimized to minimize the hardware costs. This is especially true for mass-market products such as household appliances and other consumer products. In these markets, margins are often thin, and minimizing hardware costs is essential for the profit margin.

However, the introduction of smartphones has started to challenge this approach. Smartphone revenues and profits are now driven to a large extent by apps delivered through app stores. Smartphones are often equipped with new capabilities such as extra sensors, which have no concrete use cases upon release of the new hardware. Instead, manufacturers are betting on the ingenuity of the external developer community to make use of these new capabilities and deliver additional, shared revenue via apps. This means that the revenue and profit perspective is not limited to the initial phone sales; instead, this is looked at through the lens of the total lifetime value.

The same holds true for some car manufacturers: Instead of minimizing the cost for the car BOM, they invest more in advanced hardware, even if this hardware is not fully utilized by the software in the beginning. Utilizing Over-the-Air capabilities, OEMs are constantly optimizing and extending the software that uses advanced hardware capabilities.

Of course this can be a huge bet, and it is not always clear whether it will pay off. Take, for example, a smart kitchen appliance. Instead of building it according to a minimal spec, one can provide a more generous hardware spec, including additional sensors (cooking temperature, weight, volume, etc.), which might only be fully utilized after the Start of Production of the hardware – either by providing a partner app store or even a fully open app store. In the early stages of such new product development, this can be a risk if there is no proof point that partners will jump on board – but on the other hand, the upside can be significant (Fig. 21.34).

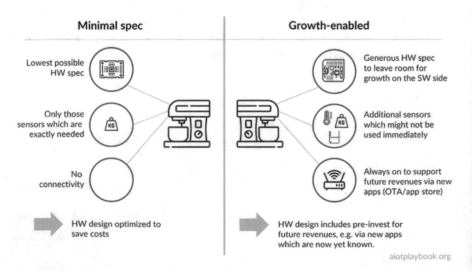

Fig. 21.34 Minimizing costs vs. planning for digital growth

21.5.4 Managing System Evolution

One of the biggest challenges for successful products with multiple system components is managing the evolution of the system design and its components over time.
This is already true for the software/AI side, especially from a configuration and
version management point of view. However, at least here, we can apply as many
changes as needed, even after the SOP (assuming OTA is enabled for all edge components). This is much more difficult for hardware since hardware upgrades are
significantly costlier than software upgrades. The following looks at some examples
to discuss this in more detail.

Example 1: Smartphones
Since the release of the first iPhone in 2007, the smartphone industry has constantly
enhanced their offerings, releasing many new versions of phone hardware, phone
OS upgrades, core application upgrades, and updates for cloud-based backend services. Backward compatibility is a key concern here: smartphone manufacturers are
interested in continuously evolving and optimizing their offerings. However, it is
not always possible to ensure backwards compatibility for every change – both on
the software as well as the hardware side – because managing too many variants
simply increases complexity to a level where it becomes unmanageable. A key prerequisite for managing compatibility across different hardware versions often boils
down to creating and maintaining standardized interfaces. Examples include the
interfaces between smartphones and headsets, or smartphones and chargers. How
open these interfaces are is another topic for debate – some vendors prefer closed
ecosystems.

Jan Bosch is Professor at Chalmers University and Director of the Software
Center: *Today, all the different hardware and software components of the smartphone ecosystem are intrinsically interwoven, forming an integrated digital offering. I usually don't care about the individual bits or atoms anymore. I am buying
into the digital offering as a whole. And this should always be available to me in
the most current version. This means frequent and proactive software upgrades,
as well as periodic upgrades of electronics and hardware. I get new, phone every
one or two years, and I don't even notice the difference anymore. OK, the camera
is a little bit better, but basically it's the same thing, right? And this is a good
thing. Because I am getting the value from the digital offering, and I don't care
how they are handling the mechanics and electronics and the software and the
AI. I just want the offering, and I am paying for that. Would it be better if I could
replace only parts of the phone like the battery or the main board? Yes, but in the
greater scheme of things it is working for me. I am not looking at my smartphone
as a physical offering anymore. It's the digital end-to-end offering that I have
bought into* (Fig. 21.35).

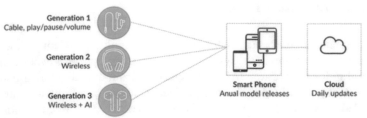

Fig. 21.35 Hardware evolution

Example 2: Electric Vehicles and Automated Driving

Another interesting example of system evolution are modern electric vehicles (EVs), and especially their Driver Assistance (DA) or Automated Driving (AD) systems. Early movers such as Tesla are constantly evolving and optimizing their products. Tesla has even gone as far as to develop their own chips and computers both for the on-board computers, and for the backend AI-training platforms.

In 2019, Tesla introduced their "Hardware 3.0" or "FSD Computer" (Full Self-Driving). This is a custom AI hardware, which replaces the off-the-shelf GPUs that Tesla was using until then. Tesla claims that their FSD hardware is by orders of magnitude more efficient for AI inference processing, which is needed for DA/AD functions. What is also very significant is that Tesla supports upgrades to the new hardware for customers with older cars. This requires a level of modularity and well-defined interfaces, which is quite advanced.

In 2021, Tesla unveiled its new supercomputer "Dojo", which is built entirely in-house, including the Dojo D1 chip. Dojo is optimized for training Tesla's advanced neural networks to support their self-driving technology. Together, these are significant, multibillion investments in creating a deeply integrated AI company (Fig. 21.36).

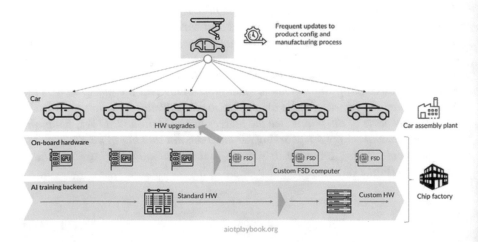

Fig. 21.36 Modern EV HW evolution

Jan Bosch from Chalmers University and the Software Center: *With cars effectively becoming digital/physical products, we are seeing much more fluidity in this
space, to match the fast advancements in technology development. Some of the leading EV manufacturers are taking a very different approach here. For them, the car
is constantly evolving. The car manufactured in July is an updated version of the car
manufactured in June. And again, the same in August. Taking this to the extreme,
they might have two floors on their factory: One floor, they are manufacturing the
cars according to the latest spec. At the floor below, they are constantly tweaking
and twisting and doing all kinds of improvements to the car architecture. And when-
ever they are satisfied with the improvements, they bring it up to the manufacturing
floor to get the update into production. This means that the next version of the car
will be manufactured with the next version of the mechanics or electronics, or whatever it was they have optimized. This might not be a reality for most of the incumbent OEMs today. There are of course questions regarding functional safety and
homologation. However, if you look at the market valuation of some of these more
agile OEMs, it seems clear that this is where the world is going.*

*This is also a general mindset thing. Instead of long-term planning where every-
thing is cast in stone for a longer period, we need to look at this as a flow. This relates
to manufacturing but also to procurement. We need more flexible contracts which
support this. Instead of focusing on getting the best possible deal for the next 100,000
sensors, I need a contract to get a flow of sensors, which I can change at any point in
time. If I then decide to go from one sensor to another or from one hardware board
to another, I can do this, because everything is set up as a flow system – enabling fast
and rapid change. This is no longer about the upfront cost for the Bill of Materials.
This is about the lifetime value that I can create from that system.*

*You can categorize the companies I am working with into two buckets. The first
bucket includes the companies that do not want any improvements. They just want
the system to continue to work as is. So all you do here is bug fixing in the beginning,
with feedback from early version in the field. But once this is stabilized, you do not
change the running system. The second bucket includes the companies that are looking for continuous product improvements. For these customers, the most important
rule is 'Thou shall be on the latest and greatest version at any point in time'. So you
do not get to choose between hardware version 17.14, software version 18.15, and
machine learning model version 19.16. No, you will always have the latest version.
That is the only way you can manage the complexity here. Let's say you are doing
two hardware platform updates a year. This means that in a two-year period, you
have four versions of the hardware. Customers might obtain a grace period of six
months before they are forced to upgrade. This means you will always only have two
or three versions of your hardware platforms that you are supporting at any point in
time. You must take a very proactive approach to limiting the variance space.
Otherwise, the complexity will kill you.*

*Of course you must find a suitable model with your customers. For example,
many of the car manufacturers that I am working with are initially selling their new
cars via a leasing model for two to three years. After this, the cars are sold to the
private market. This is a very typical pattern. What if I could provide an electronics*

package upgrade after the initial leading period, and thus extend the lifetime of the car and increase the value for the next owner. Tesla did exactly this with their upgrade option to FSD 3.0, which they are offering to owners of older cars. In the future, being able to not only upgrade software and AI but also entire sub-systems, including hardware, will be a key differentiator.

Chapter 22
AIoT Product/Solution Design

Dirk Slama

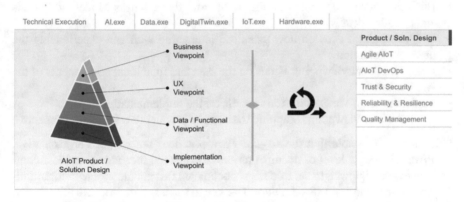

Fig. 22.1 Product/Solution design

The idea of a more detailed design document may seem old fashioned to someone who is used to working in small, agile development teams. After all, the Agile manifesto itself values *working software over comprehensive documentation* and emphasizes *the most efficient and effective method of conveying information to and within a development team is face-to-face conversation*. However, in large-scale, multi-team, multisite projects, a certain amount of documentation is required to ensure that all teams and stakeholders are aligned and working in synch. Working across organizational boundaries will add to the need for more detailed documentation of requirements and design decisions. Finally, some types of procurement contracts will require detailed specifications and SLAs (see Sourcing_and_Procurement).

D. Slama (✉)
Ferdinand Steinbeis Institute, Berlin, Germany
e-mail: dirk.slama@bosch.com

Given that an AIoT-enabled product or solution will contain different building blocks, such as AI, hardware, software, embedded components, etc., it is likely that it will face many of these constraints. Consequently, the *Digital Playbook* proposes to create and maintain a product/solution architecture that captures key requirements and design decisions in a consistent manner (Fig. 22.1).

22.1 AIoT Design Viewpoints and Templates

To provide a consistent and comprehensive design for AIoT-enabled products or solutions, the *Digital Playbook* proposes a set of design viewpoints, each with a specific set of design templates:

- Business Viewpoint: Builds on the input from the Business Model, adds KPIs and planning details
- UX Viewpoint: Focus on how users are interacting with and experiencing the product or solution
- Data/Functional Viewpoint: Focus on the data and functional components of the AIoT solution
- Implementation Viewpoint: Adds details on the implementation aspects
- AIoT Product Viewpoint: Mapping to the agile product development perspective

It is important to note that the *Digital Playbook* does not propose an excessive, RUP/waterfall-style level of documentation. The general idea is to provide a comprehensive yet lightweight set of design documents that enable efficient communication between the main stakeholders. The key to success here is to keep the design documentation on a level of detail where it is meaningful but not overly complex. The agile teams must be able to apply their own mechanism to derive requirements for their backlog and provide feedback to the overarching architecture in return. As will be discussed in the following, a central story map can be a powerful tool for keeping the design decision and the individual sprint backlogs in synch.

22.2 Important Design Considerations

It is important to accept that design documentation can rarely be seen as a stable document. The waterfall approach of fixing requirements and design decisions first will not work in most cases because of the inherent complexity, volatility of requirements and too many external and internal dependencies. The agile V-Model, for example, is specifically designed to support continuous updates of the overall system design. Each v-sprint (1) in the agile v-model must return to the design and match it against the learning from the previous sprint (2). This also means that the design documentation cannot be too detailed, since otherwise it will not be possible to perform a thorough review during each sprint planning session. The design templates provided in the Digital Playbook aim to strike a pragmatic balance between comprehensiveness and manageability. The sprint backlogs for the different teams must be in synch with the overall design (3).

In most AIoT projects there will be certain parts of the project that will require a higher level of stability for the design documentation than others. This is generally true for all areas that require more long-term planning, e.g., because of procurement or manufacturing requirements, or to manage complex dependencies across organizational boundaries. This is a key problem that the solution design team must address. In many cases, this will also require some kind of architectural layering, especially from a data/functional viewpoint. In particular, the design must ensure that no stable system components have any dependencies on more volatile components. Otherwise, the changes to the volatile components will have a ripple effect on those components that are supposed to be stable (4) (Fig. 22.2).

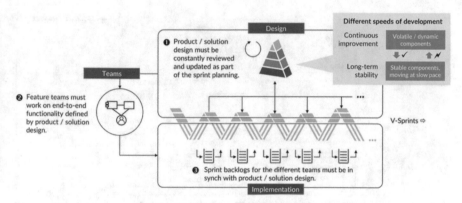

Fig. 22.2 AIoT product/solution design and the agile process

Finally, it is important to keep in mind that product/solution design and organizational structure must be closely aligned (Conway's law). The design will usually include many features that require support from different technical layers. For example, seat-heating-on-demand, which requires a smartphone component, a cloud component, and different components on the vehicle to work together to deliver this feature. The Digital Playbook is proposing to define feature teams that are working across the different technical layers. Properly deriving the setup of these feature teams from the product/solution design will be key to success (5).

22.3 ACME:Vac Example

The design section of the Digital Playbook is based on the fictitious ACME:Vac example, which is used to illustrate the different viewpoints and design templates. Robotic vacuum cleaners — or short robovacs — have become quite popular in the last decade and represent a multibillion dollar market, with a number of different players offering a variety of different models. Many of the advanced models today utilize AI to optimize robovac operations, including automatic detection of room layouts, navigation and cleaning path optimization. As such, they are a very good example of a smart, connected product provided by a Digital OEM. Differences between the design requirements of an AIoT-enabled product (Digital OEM perspective) and a solution (Digital Equipment Operator perspective) are highlighted in each section.

22.4 Business Viewpoint (Fig. 22.3)

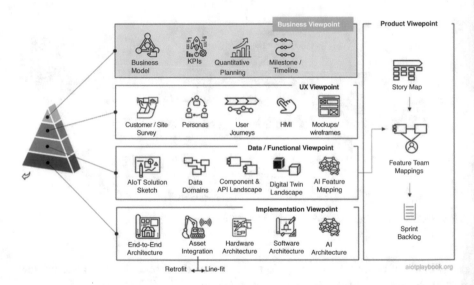

Fig. 22.3 AIoT business viewpoint

The Business Viewpoint of the AIoT Product/Solution Design builds on the different artifacts created for the Business Model. As part of the design process, the business model can be refined, e.g., through additional market research. In particular, the detailed design should include KPIs, quantitative planning, and a milestone-based timeline.

22.4.1 Business Model

The business model is usually the starting point of the product/solution design. The business model should describe the rationale of how the organization creates, delivers, and captures value by utilizing AIoT. The business model design section provides a good description of how to identify, document and validate AIoT-enabled business models. A number of different templates are provided, of which the business model canvas is the most important. The business model canvas should include a summary of the AIoT-enabled value proposition, the key customer segments to be addressed, how customer relationships are built, and the channels through which customers are serviced. Furthermore, it should provide a summary of the key activities, resources and partners required to deliver on the value proposition. Finally, a high-level summary of the business case should be provided, including cost and revenue structure (Fig. 22.4).

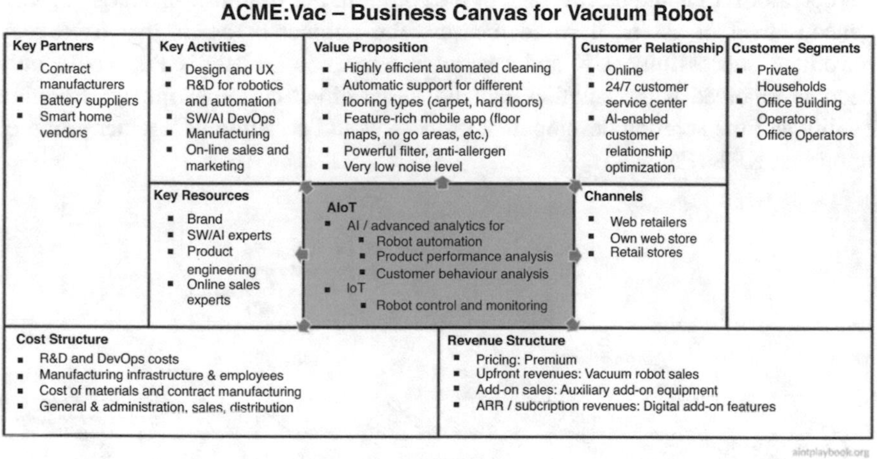

ACME:Vac – Business Canvas for Vacuum Robot

Key Partners	Key Activities	Value Proposition		Customer Relationship	Customer Segments
▪ Contract manufacturers ▪ Battery suppliers ▪ Smart home vendors	▪ Design and UX ▪ R&D for robotics and automation ▪ SW/AI DevOps ▪ Manufacturing ▪ On-line sales and marketing	▪ Highly efficient automated cleaning ▪ Automatic support for different flooring types (carpet, hard floors) ▪ Feature-rich mobile app (floor maps, no go areas, etc.) ▪ Powerful filter, anti-allergen ▪ Very low noise level		▪ Online ▪ 24/7 customer service center ▪ AI-enabled customer relationship optimization	▪ Private Households ▪ Office Building Operators ▪ Office Operators

	Key Resources
	▪ Brand
	▪ SW/AI experts
	▪ Product engineering
	▪ Online sales experts
		AIoT
		▪ AI / advanced analytics for
		  ▪ Robot automation
		  ▪ Product performance analysis
		  ▪ Customer behaviour analysis
		▪ IoT
		  ▪ Robot control and monitoring
				Channels
				▪ Web retailers
				▪ Own web store
				▪ Retail stores	

Cost Structure ▪ R&D and DevOps costs ▪ Manufacturing infrastructure & employees ▪ Cost of materials and contract manufacturing ▪ General & administration, sales, distribution			**Revenue Structure** ▪ Pricing: Premium ▪ Upfront revenues: Vacuum robot sales ▪ Add-on sales: Auxiliary add-on equipment ▪ ARR / subcription revenues: Digital add-on features		

aiotplaybook.org

Fig. 22.4 ACME:Vac business model canvas

The fictitious ACME:Vac business model assumes that AI and IoT are used to enable a high-end vacuum cleaning robot, which will be offered as a premium product (not an easy decision - some argue that the mid-range position in this market is more attractive). AI will be used not only for robot control and automation but also for product performance analysis, as well as analysis of customer behaviour. This intelligence will be used to optimize the customer experience, create customer loyalty, and identify up-selling opportunities.

22.4.2   Key Performance Indicators

Many organizations use Key Performance Indicators (KPIs) to measure how effectively a company is achieving its key business objectives. KPIs are often used on multiple levels, from high-level business objectives to lower-level process or product-related KPIs. In our context, the KPIs would either be related to an AIoT-enabled product or solution.

A Digital OEM that takes a smart, connected product to market usually has KPIs that cover business performance, user experience and customer satisfaction, product quality, and the effectiveness and efficiency of the product development process.

A Digital Equipment Operator who is launching a smart, connected solution to manage a particular process or a fleet of assets would usually have solution KPIs that cover the impact of the AIoT-enabled solution on the business process that it is supporting. Alternatively, business-related KPIs could measure the performance of the fleet of assets and the impact of the solution on that performance. Another typical operator KPI could be coverage of the solution. For example, in a large, heterogeneous fleet of assets, it could measure the number of assets that have been retrofitted successfully. UX and customer satisfaction-related KPIs would only become involved if the solution actually has a direct customer impact. Solution quality and the solution development process would certainly be another group of important KPIs (Fig. 22.5).

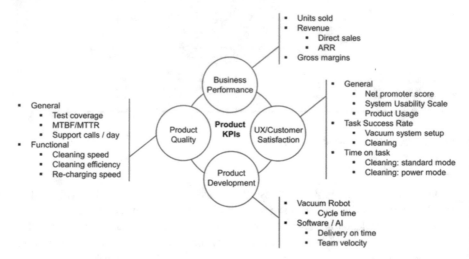

Fig. 22.5 Vacuum robot - product KPIs

The figure with KPIs shown here provides a set of example KPIs for the ACME:Vac product. The business performance-related KPIs cover the number of robovacs sold, the direct sales revenue, recurring revenue from digital add-on features, and finally the gross margin.

The UX/customer satisfaction KPIs would include some general KPIs, such as Net Promoter Score (results of a survey asking respondents to rate the likelihood that they would recommend the ACME:Vac product), System Usability Scale (assessment of perceived usability), and Product Usage (e.g., users per specific feature). The Task Success Rate KPIs may include how successful and satisfied customers are with the installation and setup of the robovac. Another important KPI in this group would measure how successful customers are actually using the robovac for its main purpose, namely, cleaning. The Time on Task KPIs could measure how long the robovac is taking for different tasks in different modes.

Product Quality KPIs need to cover a wide range of process- and product-related topics. An important KPI is test coverage. This is a very important KPI for AIoT-enabled products, since testing physical products in combination with digital features can be quite complex and expensive but a critical success factor. Incident metrics such as MTBF (mean time before failure) and MTTR (mean time to recovery, repair, respond, or resolve) need to look at the local robovac installations, as well as the shared cloud back end. Finally, the number of support calls per day can be another important indicator of product quality. Functional product quality KPIs for ACME:Vac would include cleaning speed, cleaning efficiency, and recharging speed.

Finally, the Product Development KPIs must cover all of the different development and production pipelines, including hardwire development, product manufacturing, software development, and AI development.

22.4.3 Quantitative Planning

Quantitative planning is an important input for the rest of the design exercise. For the Digital OEM, this would usually include information related to the number of products sold, as well as product usage planning data. For example, it can be important to understand how many users are likely to use a certain key feature in which frequency to be able to design the feature and its implementation and deployment accordingly.

The quantitative model for the ACME:Vac product could include, for example, some overall data related to the number of units sold. Another interesting bit of information is the expected number of support calls per year because this gives an indication for how this process must be set up. Other information of relevance for the design team includes the expected average number of rooms serviced per vacuum robot, the number of active users, the number of vacuum cleaning runs per day, and the number of vacuum cleaner bags used by the average customer per year (Fig.22.6).

Overall	Year 1	Year 2	Year 3	Year 4	Year 5
# units sold	150.000	175.000	210.000	250.000	300.000
# support calls	50.000	50.000	45.000	40.000	35.000

Per Vacuum robot	Amount	Unit
#rooms serviced	3	Total
#active users	2	Total
#vacuum runs	2	Per day
#vacuum cleaner bags	12	Per year

Fig. 22.6 Quantitative plan

For a Digital Equipment Operator, the planning data must at its core include information about the number of assets to be supported. However, it can also be important to understand certain usage patterns and their quantification. For example, a predictive maintenance solution used to monitor thousands of escalators and elevators for a railroad operator should be based on a quantitative planning model that includes some basic assumptions, not only about the number of assets to be monitored, but also about the current average failure rates. This information will be important for properly designing the predictive maintenance solution, e.g., from a scalability point of view.

22.4.4 Milestones/Timeline

Another key element of the business viewpoint is the milestone-based timeline. For the Digital OEM, this will be a high-level plan for designing, implementing and manufacturing, launching, supporting, and continuously enhancing the product.

The timeline for the ACME:Vac product differentiates between the physical product and the AIoT part (including embedded hardware and software, AI, and cloud). If custom embedded hardware is to be designed and manufactured, this could also be subsumed under the physical product workstream, depending on the organizational setup. The physical product workstream includes a product design and manufacturing engineering phase until the Start of Production (SOP). After the SOP, this workstream focuses on manufacturing. A new workstream for the next physical product generation starting after the SOP is omitted in this example. The AIoT workstream generally assumes that an AIoT DevOps model is applied consistently through all phases.

Key milestones for both the physical product and the AIoT part include the initial product design and architecture (result of sprint 0), the setup of the test lab for testing the physical product, the first end-to-end prototype combining the physical product with the AIoT-enabled digital features, the final prototype/Minimum Viable Product, and finally the SOP.

The following figure also highlights the V-Sprints, which in this example applies to both physical product development and the AIoT development. While physical product development is unlikely to deliver potentially shippable product increments at the end of each V-Sprint, it still assumes the same sprint cadence.

Because sourcing is typically such a decisive factor, the timeline includes milestones for the key sourcing contracts that must be secured. Details regarding the procurement process are omitted on this level (Fig. 22.7).

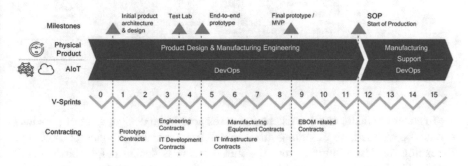

Fig. 22.7 Example milestone plan

For a Digital Equipment Operator, this plan would focus less on the development and manufacturing of the physical product. Instead, it would most likely include a dedicated workstream for managing the retrofit of the solution to the existing physical assets.

22.5 Usage Viewpoint (Fig. 22.8)

The goal of the UX (User Experience) viewpoint is to provide a holistic view of how the product or solution will be utilized by the user and other stakeholders. Good UX practice usually includes extensive product validation, including usability testing, user feedback, pilot user tests, and so on. A good starting point is usually customer surveys or interviews. In the case of an AIoT-enabled product or solution it can also make sense to include site surveys to better understand the environment of the physical products or assets.

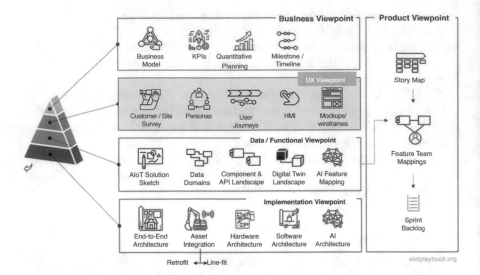

Fig. 22.8 AIoT usage viewpoint

To ensure realistic and consistent use cases across the design, a set of personas should be defined, representing the typical users of the product or solution. Revisiting the User Journey from the initial business design helps clarify many details. Finally, HMI (Human-Machine Interaction) design, early prototypes and wire frames are also essential elements of the UX viewpoint.

22.5.1 Site Surveys and Stakeholder Interviews

To capture and validate requirements, it is common practice for IT projects to perform stakeholder interviews. This should also be done in case of an AIoT product/project.

However, an AIoT project is different in that it also involves physical assets and potentially also very specific sites, e.g., a factory. Requirements can heavily depend on the type of environment in which assets are deployed. Additionally, usage patterns might vastly differ, depending on the environment. Consequently, it is highly recommended for the team responsible for the product design to spend time on-site and investigate different usage scenarios in different environments.

While many AIoT solutions might be deployed at a dedicated site, this might not be true for AIoT-enabled products. Take, for example, a smart kitchen appliance, which will be sold to private households. In this particular case, it can make sense to actually build a real kitchen as a test lab to test the usage of the product in a realistic environment. Alternatively, in the case of our Vacuum Robot, different scenarios for testing the robot must be made available, including different room types and different floor surfaces (wood panels, carpets, etc.).

22.5.2   Personas

Personas are archetypical users of the product or solution. Often, personas represent fictitious people who are based on your knowledge of real users. The UX Viewpoint should define a comprehensive set of personas that help model the product features in a way that takes the perspective of different product users into consideration. By personifying personas, the product team will ideally even develop an emotional bond to key personas, since they will accompany them through an intense development process. A persona does not necessarily need a sophisticated fictitious background story, but at least it should have a real-world first name and individual icon, as shown in Fig. 22.9.

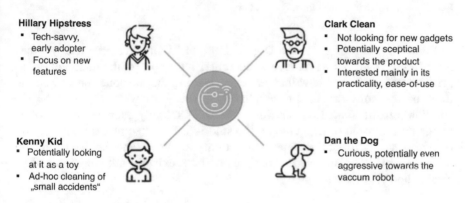

Hillary Hipstress
- Tech-savvy, early adopter
- Focus on new features

Clark Clean
- Not looking for new gadgets
- Potentially sceptical towards the product
- Interested mainly in its practicality, ease-of-use

Kenny Kid
- Potentially looking at it as a toy
- Ad-hoc cleaning of „small accidents"

Dan the Dog
- Curious, potentially even aggressive towards the vaccum robot

Fig. 22.9 AIoT personas

22.5.3   User Journeys

The initial User Journeys from the Business Model design phase can be used as a starting point. Often, it can be a good idea in this phase of the product design to create individual journey maps for different scenarios, adding more detail to the original, high-level journey.

The example user journey for ACME:Vac shown here is not that different from most user journey designs found for normal software projects. The main difference is that the user journey here is created along the life cycle of the product from the customer's point of view. This includes important phases such as Asset Activation, Asset Usage and Service Incidents (Fig. 22.10).

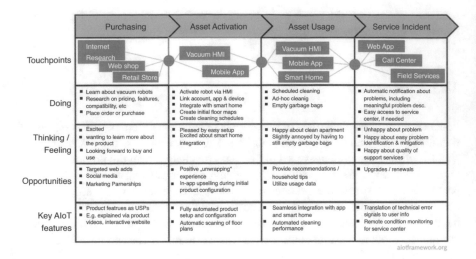

	Purchasing	Asset Activation	Asset Usage	Service Incident
Touchpoints	Internet Research / Web shop / Retail Store	Vacuum HMI / Mobile App	Vacuum HMI / Mobile App / Smart Home	Web App / Call Center / Field Services
Doing	▪ Learn about vacuum robots ▪ Research on pricing, features, compatibility, etc ▪ Place order or purchase	▪ Activate robot via HMI ▪ Link account, app & device ▪ Integrate with smart home ▪ Create initial floor maps ▪ Create cleaning schedules	▪ Scheduled cleaning ▪ Ad-hoc cleanig ▪ Empty garbage bags	▪ Automatic notification about problems, including meaningful problem desc. ▪ Easy access to service center, if needed
Thinking / Feeling	▪ Excited ▪ wanting to learn more about the product ▪ Looking forward to buy and use	▪ Pleased by easy setup ▪ Excited about smart home integration	▪ Happy about clean apartment ▪ Slightly annoyed by having to still empty garbage bags	▪ Unhappy about problem ▪ Happy about easy problem identification & mitigation ▪ Happy about quality of support services
Opportunities	▪ Targeted web adds ▪ Social media ▪ Marketing Parnerships	▪ Positive „unwrapping" experience ▪ In-app upselling during initial product configuration	▪ Provide recommendations / household tips ▪ Utilize usage data	▪ Upgrades / renewals
Key AIoT features	▪ Product featrues as USPs ▪ E.g. explained via product videos, interactive website	▪ Fully automated product setup and configuration ▪ Automatic scaning of floor plans	▪ Seamless integration with app and smart home ▪ Automated cleaning performance	▪ Translation of technical error signals to user info ▪ Remote condition monitoring for service center

aiotframework.org

Fig. 22.10 Customer journey for vacuum robot

From the point of view of a Digital Equipment Operator, the user journey most likely focuses less on an end customer but more on the different enterprise stakeholders and how they are experiencing the introduction and operations of the solution. Important phases in the journey here would be the solution retrofit, standard operations, and what actually happens in case of an incident monitored or triggered by the solution. For example, for a predictive maintenance solution, it is important not only to understand the deep algorithmic side of it but also how it integrates with an existing organization and its established processes.

22.5.4   UX/HMI Strategy

The UX/HMI strategy will have a huge impact on usability. Another important factor is the question of how much the supplier will be able to learn about how the user is interacting with the product. This is important, for example, for product improvements but also potentially for upselling and digital add-on services (Fig. 22.11).

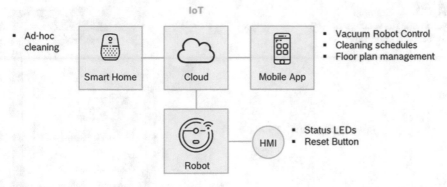

Fig. 22.11 UX/HMI for vacuum robot

The HMI strategy for ACME:Vac seems relatively straightforward at first sight: HMI features on the robovac are reduced to a minimum, including only some status LEDs and a reset button. Instead, almost all of the user interaction is done via the smartphone app. In addition, some basic commands such as starting an ad hoc cleaning run are supported via smart home integration.

It is important that the decision for the HMI strategy will have a huge impact not only on usability and customer experience but also on many other aspects, such as product evolvability (a physical HMI cannot be easily updated, while an app-based HMI can), customer intimacy (easier to learn how a customer is using the product via digital platforms), as well as the entire design and development process, including manufacturing (none needed for app-based HMI).

However, the risk of completely removing the HMI from the physical product should also not be underestimated. For example, in the case of bad connectivity or unavailability of a required cloud backend, the entire physical product might become entirely unusable.

22.5.5   *Mockups/Wireframes/Prototypes*

To ensure a good user experience, it is vital to try out and validate different design proposals as early as possible. For purely software-based HMI, UI mockups or wireframes are a powerful way of communicating the interactive parts of the product design, e.g., web interfaces and apps for smartphones or tablets. They should initially be kept on the conceptual level. Tools such as Balsamiq offer a comic-style way of creating mockups, ensuring that they are not mistaken for detailed UI implementation designs. The figure shown here provides a mockup for the ACME:Vac floor map management feature on the ACME:Vac app based on this style (Fig. 22.12).

Fig. 22.12 Example
wireframe for vacuum
robot smart phone app

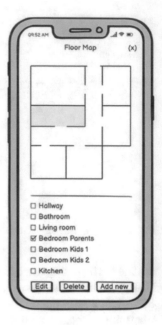

It should be noted that the validation of the UX for physical HMI can require the actual creation of a physical prototype. Again, this should be done as early as possible in the design and development phase, because any UX issued identified in the early stages will help save money and effort further downstream. For example, while the HMI on board the ACME:Vac robot is kept to a minimum, there are still interesting aspects to be tested, including docking with the charging station and replacement of the garbage bag.

22.6   Data/Functional Viewpoint (Fig. 22.13)

The Data and Functional Viewpoint provides design details that focus on the overall functionality of the product or solution, as well as the underlying data architecture. The starting point can be a refinement of the AIoT Solution Sketch. A better understanding of the data architecture can be achieved with a basic data domain model. The component and API landscape is the initial functional decomposition, and will have a huge impact on the implementation. The optional Digital Twin landscape helps understand how Digital Twin as a concept fit into the overall design. Finally, AI Feature Mapping helps identify which features are best suited to an implementation with AI.

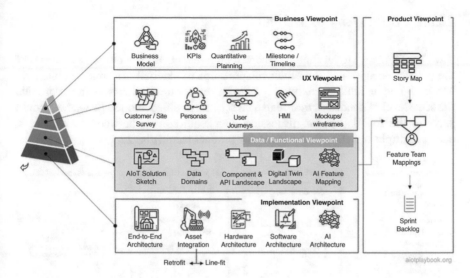

Fig. 22.13 AIoT Data/Functional viewpoint

22.6.1   AIoT Solution Sketch

The AIoT Solution Sketch from the Business Model can be refined in this perspective, adding more layers of detail, in a slightly more structured process of presentation. The solution sketch should include the physical asset or project (the robovac in our example), as well as other key assets (e.g., the robovac charging station) and key users. Since interactions between the physical assets and the back end are key, they should be listed explicitly. The sketch should also include an overview of key UIs, the key business processes supported, the key AI and analytics-related elements, the main data domains, and external databases or applications (Fig. 22.14).

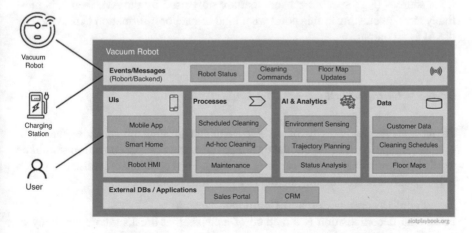

Fig. 22.14 Solution sketch for vacuum robot

22.6.2 Data Domain Model

The Data Domain Model should provide a high-level overview of the key entities of the product design, including their relationships to external systems. The Domain Model should include approximately a dozen key entities. It does not aim to provide the same level of detail as a thorough data schema or object model. Instead, it should serve as the foundation for discussing data requirements between stakeholders, across multiple stakeholder groups in the product team (Fig. 22.15).

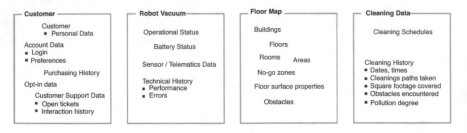

Fig. 22.15 Data domain model for vacuum robot

For example, the main data domains that have been identified for the ACME:Vac product are the customer, the robovac itself, floor maps and cleaning data. Each of these domains is described by listing the 5–10 key entities within. This is typically a good level of detail: sufficiently meaningful for planning discussions, without getting lost in detail.

The design team must make a decision on whether the required data from an AI perspective should already be included here. This can make sense if AI-related data also play an important role in other, non AI-based parts of the system. In this case, potential dependencies can be identified and discussed here. If the AI has dedicated input sources (e.g., specific sensors that are only used by the AI), then it is most likely more interesting at this point what kind of data or information is provided by the AI as an output.

22.6.3 Component and API Landscape

To manage the complexity of an AIoT-enabled product or solution, the well-established approach of functional decomposition and componentization should be applied. The results should be captured in a high-level component landscape, which helps visualize and communicate the key functional components.

Functional Decomposition and Componentization
Functional decomposition is a method of analysis that breaks a complex body of work down into smaller, more easily manageable units. This "Divide & Conquer"

strategy is essential for managing complexity. Especially if the body of work cannot be implemented by a single team, splitting the work in a way that it can be assigned to different teams in a meaningful way becomes very important. Since a system's architectural structure tends to be a mirror image of the organizational structure, it is important that team building follows the functional decomposition process. The idea of "feature teams" to support this is discussed in the AIoT Product Viewpoint.

Another key point of functional decomposition is functional scale: without effective functional decomposition and management of functional dependencies, it will be difficult to build a functionally rich application. It does not stop at building the initial release. Most modern software-based systems are built on the agile philosophy of continuous evolution. While an AIoT-enabled product or solution does not only consist of software — it also includes AI and hardware — enabling evolvability is usually a key requirement. Functional decomposition and componentization will enable the encapsulation of changes and thus support efficient system evolution.

The logical construct for encapsulating key data and functionality in an AIoT system should be the component. From a functional viewpoint, components are logical constructs, independent of a particular technology. Later in the implementation viewpoint, they can be mapped to specific programming languages, AI platforms, or even specific functionality implemented in hardware. Additionally, component frameworks such as microservices can be added where applicable (Fig. 22.16).

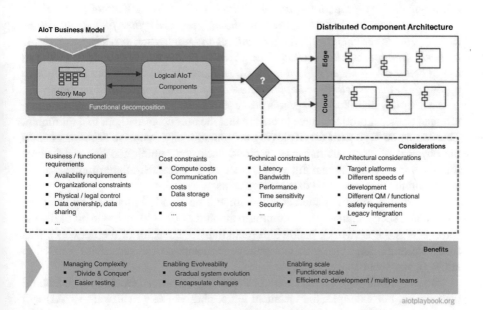

Fig. 22.16 Functional decomposition and componentization

The functional decomposition process should go hand in hand with the development of the agile story map (see AIoT Product Viewpoint) since the story map will contain the official definition of the body of work, broken down into epics and features.

The first iteration of the component landscape can actually be very close to the story map, since it should truly only focus on the functional decomposition. In a second iteration, the logical AIoT components must be mapped to a distributed component architecture. This perspective is actually somewhat between the functional and implementation viewpoints. The mapping to the distributed component architecture must take a number of different aspects into consideration, including business/functional requirements, cost constraints, technical constraints and architectural constraints.

A key functional requirement simply is availability. In a distributed system, remote access to a component always has a higher risk of the component not being available, e.g., due to connectivity issues. Other business-driven aspects are organizational constraints (especially if different parts of the distributed system are developed by different organizational units), physical control and legal aspects (deploying critical data or functionality in the field or in certain countries can be difficult). This is also closely related to data ownership and data sharing requirements.

Achim Nonnenmacher, expert for Software-defined Vehicle at Bosch comments: *Because of the availability issues related to distributed applications, many leading services are using a capability-based architecture. For example, many smart phones have two versions of their key services - one which works with the data and capabilities available on the phone, and one which works only with cloud connectivity. For example, you can say "Hey, bring me home", and the offline phone will still be able to provide rudimentary voice recognition and navigation services using the AI and map data on the phone. Only if the phone is online will it be able to make full use of better cloud-based AI and data. We still have to learn in many ways how to apply this to the vehicle-based applications of the future, but this will be important.*

Another key distribution aspect is cost constraints. For example, many manufacturers of physical products have to ensure that the unit costs are kept to a minimum. This can be an argument for deploying computationally intensive functions not on the physical product but rather on a cloud back end, which can better distribute loads coming from a large number of connected products. Similar arguments apply to communication costs and data storage costs.

Furthermore, the distributed architecture will be greatly influenced by technical constraints, such as latency (the minimum amount of time for a single bit of data to travel across the network), bandwidth (data transfer capacity of the network), performance (e.g., edge vs. cloud compute performance), time sensitivity (related to latency), and security.

Finally, a number of general aspects should be considered from the architectural perspective. For example, the technical target platform (e.g., software vs. AI) will play a role. Another factor is the different speed of development: a single component should not combine functionalities that will evolve at significantly different speeds. Similarly, one should avoid combining functionality, which only requires standard Quality Management with functionality, which must be treated as functional safety relevant to a single component. In this case, the QM functionality must also be treated as functional safety relevant, making it costliest to test, maintain and update.

While some of these constraints and considerations are of a more technical nature, they need to be combined with more business or functional considerations when designing the distributed component architecture.

Component Landscape

The result of the functional decomposition process should be a component landscape, which focuses on functional aspects but already includes a high-level distribution perspective.

The example shown in Fig. 22.17 is the high-level component landscape for the ACME:Vac product. This component landscape has three swimlanes: one for the robovac (i.e., the edge platform), one for the main cloud service, and one for the smartphone running the ACME:Vac mobile app. The components of the robot include basic robot control and sensor access, as well as the route/trajectory calculation. These components would most likely be based on an embedded platform, but this level of detail is omitted from a functional viewpoint. In addition, the robot will have a configuration component, as well as a component offering remote services that can be accessed from the robot control component in the cloud. In addition, the cloud contains components for robot configuration, user configuration management, map data, and the management of the system status and usage history. Finally, the mobile app has a component to manage the main app screen, map management, and remote robot configuration.

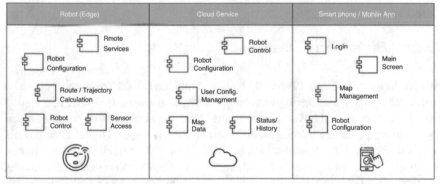

Fig. 22.17 Example: initial component architecture

API Management

In his famous "API Mandate", Jeff Bezos — CEO of Amazon at the time — declared that *"All teams will henceforth expose their data and functionality through service interfaces."* at Amazon. If the CEO of a major companies gets involved on this level, you can tell how important this topic is.

APIs (Application Programming Interfaces) are how components make data and functionality available to other components. Today, a common approach is so-called RESTful APIs, which utilize the popular HTTP internet protocol. However, there are many different types of APIs. In AIoT, another important category of APIs is between the software and the hardware layer. These APIs are often provided as low-level c APIs (of course, any c API can again be wrapped in a REST API and exposed to remote clients). Most APIs support a basic request/response pattern to enable interactions between components. Some applications require a more message-oriented, de-coupled way of interaction. This requires a special kind of API.

Regardless of the technical nature of the API, it is good practice to document APIs via an API contract. This contract defines the input and output arguments, as well as the expected behavior. "Interface first" is an important design approach which mandates that before implementing a new application component, one should first define the APIs, including the API contract. This approach ensures de-coupling between component users and component implementers, which in turn reduces dependencies and helps managing complexity. Because APIs are such an important part of modern, distributed system development, they should be managed and documented as key artefacts, e.g. using modern API management tools which support API documentation standards like OpenAPI.

From the system design point of view, the component landscape introduced earlier should be augmented with information about the key APIs supported by the different components. For example, the component landscape documentation can support links to the detailed API documentations in different repositories. This way, the component landscape provides a high level description not only of how data and functionality is clustered in the system, but also how it can be accessed.

22.6.4 Digital Twin Landscape

As introduced in the Digital Twin 101 section, using the Digital Twins concept can be useful, especially when dealing with complex physical assets. In this case, a Digital Twin Landscape should be included with the Data/Functional Viewpoint. The Digital Twin Landscape should provide an overview of the key logical Digital Twin models and their relationships. Relationships between Digital Twin model elements can be manifold. They should be used to help define the so-called ""knowledge graph""across different, often heterogeneous data sources used to construct the Digital Twin model.

In some cases, the implementation of the Digital Twin will rely on specific standards and Digital Twin platforms. The Digital Twin Landscape should keep this in mind and only use modeling techniques that will be supported later by the implementation environment. For example, the Digital Twins Definition Language (DTDL) is an open standard supported by Microsoft, specifically designed to support modeling of Digital Twins. Some of the rich features of DTDL include Digital Twin interfaces and components, different kinds of relationships, as well as persistent properties and transient telemetry events.

In the example shown in Fig. 22.18, these modeling features are used to create a visual Digital Twin Landscape for the ACME:Vac example. The example differentiates between two types of Digital Twin model elements: system (green) and environment (blue). The system elements relate to the physical components of the robovac system. Environment elements relate to the environment in which a robovac system is actually deployed.

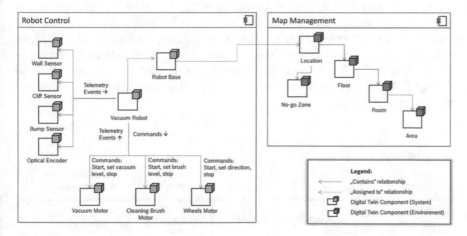

Fig. 22.18 Digital twin landscape

This differentiation is important for a number of reasons. First, the Digital Twin system elements are known in advance, while the Digital Twin environment elements actually need to be created from sensor data (see the discussion on Digital Twin reconstruction).

Second, while the Digital Twin model is supposed to provide a high level of abstraction, it cannot be seen completely in isolation of all the different constraints discussed in the previous section. For example, not all telemetry events used on the robot will be directly visible in the cloud. Otherwise, too much traffic between the robots and the cloud will be created.

This is why the Digital Twin landscape in this example assigns different types of model elements to different components. In this way, the distributed nature of the component landscape is taken into consideration, allowing for the creation of realistic mapping to a technical implementation later on.

22.6.5 AI Feature Mapping

The final element in the Data/Functional Viewpoint should be an assessment of the key features with respect to suitability for implementation with AI. As stated in the introduction, a key decision for product managers in the context of AIoT will be whether a new feature should be implemented using AI, Software, or Hardware. To ensure that the potential for the use of AI in the system is neither neglected nor overstated, a structured process should be applied to evaluate each key feature in this respect.

In the example shown in Fig. 22.19, the features from the agile story map (see AIoT Product Viewpoint) are used as the starting point. For each feature, the expected outcome is examined. Furthermore, from an AI point of view, it needs to be understood which live data can be made available to potentially AI-enabled components, as well as which training data. Depending on this information, an initial recommendation regarding the suitability of a given feature for implementation with AI can be derived. This information can be mapped back to the overall component landscape, as indicted by Fig. 22.20. Note also that a new component for cloud-based model training is added to this version of the component landscape. Not that this level of detail does not describe, for example, details of the algorithms used, e.g. Simultaneous Localization and Mapping (SLAM), etc.

Product Feature	Expected Outcome	Required live input data	Required training data	Implementation with AI recommended?
Route calculation	Optimized route, ad-hoc trajectory changes	Real-time sensor data	Data from cleaning runs in test lab	✓
Floor plan creation	Floor plan (created automatically)	Historic data from cleaning runs at customer	Data from cleaning runs in test lab	✓
...	...	...	...	...

Fig. 22.19 AI feature mapping

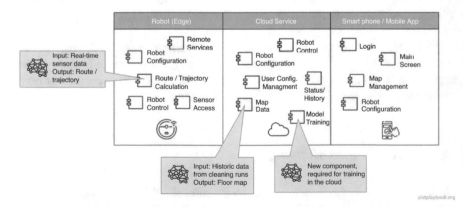

Fig. 22.20 Identifying AI-enabled components

22.7   Implementation Viewpoint (Fig. 22.21)

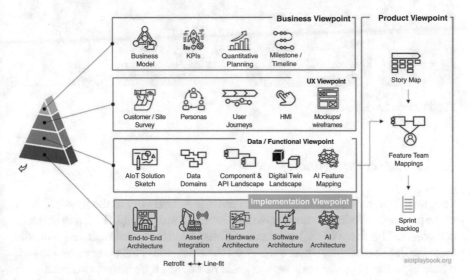

Fig. 22.21   AIoT implementation viewpoint

The Implementation Viewpoint must provide sufficient detail to have meaningful technical discussions between the different technical stakeholders of the product team. However, most design artifacts in this viewpoint will still be on a level of abstraction which will hide many of the different details required by the implementation teams. Nevertheless, it is important to find a common language and understanding between the different stakeholders, including a realistic mapping to the Data/Functional Viewpoint.

The AIoT Implementation Viewpoint should at least include an End-to-End Architecture, details on the planned integration with the physical asset (either following a line-fit or retrofit approach), as well as high-level hardware, software and AI architectures.

22.7.1   End-to-End Architecture

The End-to-End Architecture should include the integration of physical assets, as well as the integration of existing enterprise applications in the back end. In between, an AIoT system will usually have edge and cloud or on-premises back end components. These should also be described with some level of detail, including technical platforms, middleware, AI and Digital Twin components, and finally the business logic itself (Fig. 22.22).

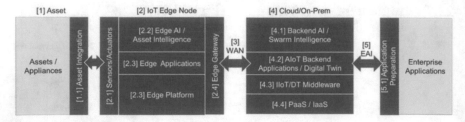

Fig. 22.22 IoT architecture

22.7.2 Asset Integration

The Asset Integration perspective should provide an overview of the physical parts of the product, including sensors, antennas, battery/power supply, HMI, and onboard computers. The focus is on how these different elements are integrated with the asset itself. For example, where exactly on the asset would the antenna be located, where to position key elements such as main board, battery, sensors, etc. Finally, an important question will concern wiring for power supply, as well as access to local bus systems (Fig. 22.23).

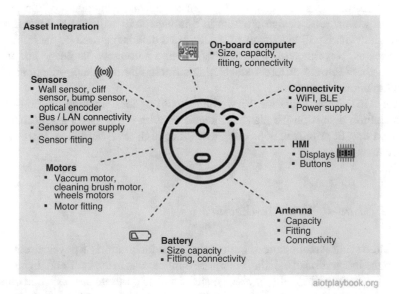

Fig. 22.23 Asset integration

22.7.3 Hardware Architecture

Depending on the requirements of the AIoT system, custom hardware development can be an important success factor. The complexity of custom hardware design and development should not be underestimated. From the hardware design point of view, a key artefact is usually the schematic design of the required PCBs (Printed Circuit Boards).

The ACME:Vac example shown in Fig. 22.24 includes the main control unit, HMI, power management, sensors, wireless connectivity, signal conditioning, and finally the control of the different motors.

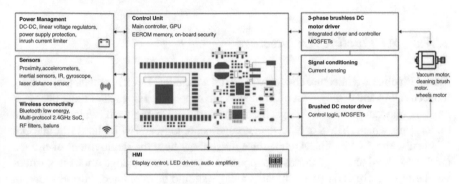

Fig. 22.24 Robovac hardware architecture

22.7.4 Software Architecture

The technical software architecture should have a logical layering, showing key software components and their main dependencies. For the ACME:Vac example, the software architecture would include two main perspectives: the software architecture on the robovac (Fig. 22.25) and the backend architecture (not shown).

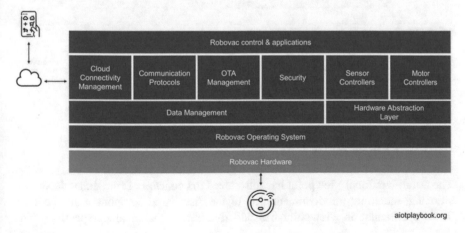

Fig. 22.25 Example: Robovac SW architecture

Depending on the type of organization, software architecture will be ad hoc or follow standards such as the OpenGroup's [1] framework. TOGAF, for example, provides the concept of Architecture and Solution Building Blocks (ABB and SBB, respectively), which can be useful in more complex AIoT projects.

The example shown here is generic (like an ABB in TOGAF terms). Not shown here is a mapping of the software architecture to concrete products and standards (like a TOGAF SBB), which would usually be the case in any project. However, the *Digital Playbook* does not want to favor any particular vendor and is consequently leaving this exercise to the reader.

22.7.5   AI Pipeline Architecture

The AI Pipeline Architecture should explain, on a technical level, how data preparation, model training and deployment of AI models are supported. For each of these phases, it must be understood which AI-specific frameworks are being used, which additional middleware, which DBMS or other data storage technology, and which hardware and OS.

Finally, the AI Pipeline Architecture must show how the deployment of trained models to cloud and edge nodes is supported. For distributed edge nodes in particular, the support for OTA (over-the-air) updates should be explained. Furthermore, in the case of AI on distributed edge nodes, the architecture must explain how model monitoring data are captured and consolidated back in the cloud (Fig. 22.26).

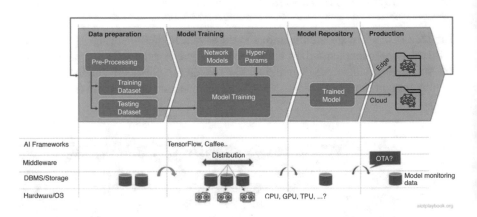

Fig. 22.26 AI architecture

22.7.6   Putting It All Together

The Data/Functional Viewpoint has introduced the concept of functional decompositioning, including the documentation of the distributed component architecture. The Implementation Viewpoint has added different technical perspectives. The different functional components must be mapped to technology-specific pipelines. For this, feature teams must be defined that combine the required technical skills/ access to the required technical pipelines for a specific feature (see the AIoT Product Viewpoint for a more detailed discussion on feature teams and how they are assigned) (Fig. 22.27).

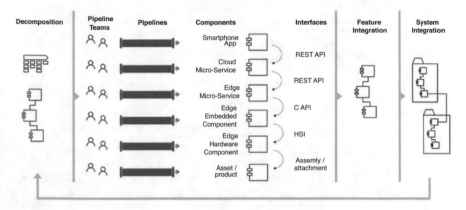

Fig. 22.27 Architectural decomposition and system integration

   The results from the different technical pipelines are individual technical components that must be integrated via different types of interfaces. For example, smartphone, cloud and edge components can be integrated via REST interfaces. On the edge, embedded components are often integrated via C interfaces. The integration between embedded software and hardware is done via different types of Hardware/ Software Interfaces (HSI). Finally, any AIoT hardware components must be physically integrated with the actual physical product. During the development/testing phase, this will usually be a manual process, while later it will be either a standardized retrofit or line-fit process.

   All of this will be required to integrate the different components required for a specific feature across the different technical pipelines. Multiple features will be integrated to form the entire system (or system-of-systems, depending on the complexity or our product or solution).

22.8   Product Viewpoint (Fig. 22.28)

The Product Viewpoint must map the other elements of the Product Architecture to the key elements of an agile product organization. The main artefact here is the agile story map, which is the highest level structural description of the entire body of work. Feature team mapping supports the mapping of the work described in the story map to the teams needed to implement the different product features. Finally, for each team and each sprint an individual sprint backlog must be created based on the story map and the results of the feature team mappings.

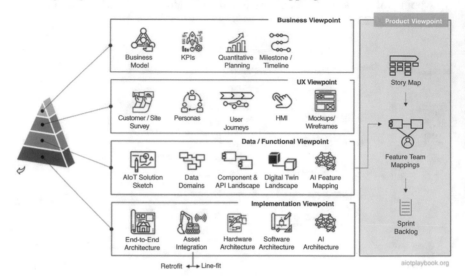

Fig. 22.28 AIoT product viewpoint

22.8.1   Story Map

It is best practice in the agile community to breakdown a larger body of work into specific work items using a hierarchical approach. Depending on the method applied, this hierarchy could include themes, epics, features, and user stories.

A story map organizes user stories in a logical way to present the big picture of the product. Story maps help ensure that user stories are well balanced, covering all important aspects of the planned solution at a similar level of detail. Story maps provide a two-dimensional graphical visualization of the Product Backlog. Many modern development support tools (such as Jira) support automatic visualization of the product backlog as a story map.

The AIoT Framework assumes the following hierarchy:

- Epic: A high-level work description, usually outlining a particular usage scenario from the perspective of one of multiple personas
- Feature: A specific feature to support an epic
- User Story: short requirements written from the perspective of an end user

Depending on the complexity of the project and the agile method chosen, this may need to be adapted, e.g. by further adding *themes* as a way of bundling epics.

When starting to break down the body of work, one should first agree on a set of top-level epics, and ensure that they are consistent, do not overlap, and cover everything that is needed. For each epic, a small number of features should be defined. These features should functionally be independent (see the discussion on functional decomposition).

Finally, features can further be broken down into user stories. User stories are short and concise descriptions of the desired functionality told from the perspective of the user.

The example shown in Fig. 22.29 is the story map for the ACME:Vac product. It has six epics, including HMI, Cleaning, Maps, Navigation/Sensing, Configuration and Status/History. Each epic is broken down into a small number of key features supporting the epic. User stories are not shown on this level. Note that this story map does not include the entire mechatronic part of the system, including chassis, motor, locomotion (climbing over obstacles, etc.), home base, etc. Also, functional safety is not included here, which would be another real-world requirement.

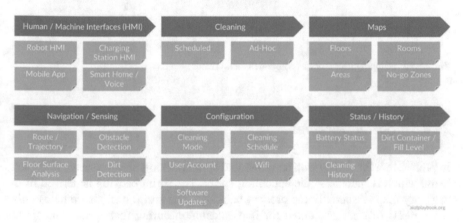

Fig. 22.29 Example: initial story map

22.8.2 Feature Team Mapping

One of the main challenges in almost all product organizations is the creation of efficient mapping between the organizational structure and the product structure (the same applies to projects and solutions). The problem here is that organizations are often more structured around skills (UX, frontend, back end, testing, etc.), while product features usually require a mixture of these skills.

Consequently, the *Digital Playbook* recommends an approach based on feature teams, which are assigned on demand to match the requirements of a specific feature. See Agile AIoT Organization for a more detailed discussion. Feature teams can exist for longer periods of time, spanning multiple sprints, if the complexity of the feature requires this.

In the example shown in Fig. 22.30, the user story "Change cleaning mode" (part of the cleaning mode configuration feature) is analyzed. The results of the analysis show that a number of components on the robovac, the cloud and mobile app must be created or extended to support this user story. A similar analysis must be done for all other user stories of the overarching feature before a proposal for the supporting feature team can be made. In this case, the feature team must include a domain expert, an embedded developer, a cloud developer, a mobile app developer, and an integration/test expert. To support the scrum approach, who in the feature team plays the role of product (or feature) owner, as well as the scrum master, must be agreed upon.

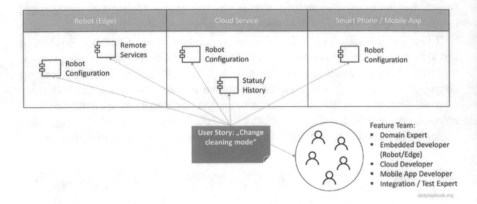

Fig. 22.30 Mapping user story to components and feature team

22.8.3 Sprint Backlogs

In preparation for each sprint, an individual sprint backlog must be created for each team, which is specific to the upcoming sprint. The sprint backlog is derived from the story map (essentially the product backlog). The sprint backlog contains only those items that are scheduled for implementation during that sprint. The sprint backlog can contain user stories to support features but also bug fixes or nonfunctional requirements.

In larger organizations with multiple feature teams, the Chief Product Owner is responsible for the overarching story map, which serves as the product backlog. He prioritizes product backlog items based on risk, business value, dependencies, size, and date needed and assigns them to the individual teams. The teams will usually refine them and create their own sprint backlogs, in alignment with the Chief Product Owner and the product/feature owners of the individual teams.

Chapter 23
Agile AIoT

Dirk Slama

Being able to address the complexity and volatility of AIoT product/solution development in an agile way is a key success factor. This chapter first takes a look at some of the key challenges to achieve this, then discusses the Agile V-Model as a way to get there, followed by a discussion of the agile AIoT organization.

23.1 Agile Versus AIoT (Fig. 23.1)

Fig. 23.1 AIoT and agility

D. Slama (✉)
Ferdinand Steinbeis Institute, Berlin, Germany
e-mail: dirk.slama@bosch.com

© The Author(s) 2023
D. Slama et al. (eds.), *The Digital Playbook*,
https://doi.org/10.1007/978-3-030-88221-1_23

23.1.1 Agility Versus AIoT: Impediments

With the emergence of Internet- and Cloud-based development, Agile software development has risen to become the de facto standard for many organizations, even though there are still many (sometimes religious) debates about how to implement Agile correctly. Software projects are usually plagued by very high levels of complexity and highly volatile requirements. Agile development is addressing this issue by combining collaborative efforts of self-organizing and cross-functional teams with a strong sense of customer-centricity and focus on value creation. Agile promotes continuous exploration, continuous integration and continuous deployment. Scrum — as probably the most prominent Agile method — is based on an adaptive planning approach, which combines higher-level epics with a more detailed requirements backlog. Detailed requirements ("user stories") are typically only created for the next 1–2 sprints ahead. Successful Scrum organizations are very thorough and consequential in their Scrum rigor while still supporting adaptive and flexible planning.

Unfortunately, in most AIoT projects, there are some impediments to a fully Agile approach, including the following:

- Cultural differences between Cloud-native developers and product/manufacturing type of engineers
- Scalability across multiple, often distributed teams (e.g. AI development, cloud and edge/embedded software development, hardware and network engineering)
- Sourcing and external dependencies which require more documentation and long-term planning
- The integration of teams working on Artificial Intelligence, which does not yet have a proven agile method to support it
- Hardware and embedded software engineering, that play by different rules than cloud-based software development
- Components/features which have to be "First Time Right", e.g., because they will be deployed on assets in the field and can no longer be easily changed afterwards
- Functional Safety requirements, which often do not play well with an agile approach

The following will look at each of these impediments in detail, followed by a recommendation for how to address them in an AIoT product organization (Fig. 23.2).

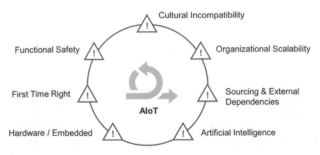

aiotplaybook.org

Fig. 23.2 Agile inhibitors

Impediment 1: Cultural Incompatibility

The first impediment for the adoption of a pure agile approach in an AIoT-driven product organization is the cultural differences typically found in heterogeneous teams that need to work together. Developers who are used to work with frequent, Cloud-enabled updates have a very different approach to project management than manufacturing-centric engineers who know that after the SOP (Start-of-Production) any change to a physical asset after it has left the factory involves a costly and painful recall procedure. As shown in the figure below, different types of organizations typically have different types of cultures - even within the same company. Applying one-size-fits all methods to such a cultural mix will be difficult (Fig. 23.3).

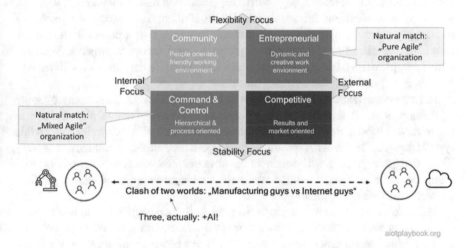

Fig. 23.3 Corporate cultures and agile setup

Depending on the culture found in the organizational units involved in the development of the AIoT product, a matching agile approach must be chosen. Approaches such as Scrum of Scrums (SoS) or Large Scale Scrum (LeSS) could be a good fit for an entrepreneurial culture, which permits a "pure" agile approach. The Scaled Agile Framework (SAFe) could be a good fit for organizations that come more from a "Command & Control" background, often combined with elements of a more matrix-like organization.

Impediment 2: Organizational Scalability

The second impediment is organizational scalability: most "pure" Agile methods such as Scrum work best for single, colocated teams. Chances are that an AIoT project will be larger, involving multiple teams that are dealing with AI development, on-asset components (Edge software, embedded, hardware), the IoT network, the backend business logic (cloud or on-premises), and the integration of existing enterprise systems. Given the diversity of these tasks, it also seems likely that the teams and team members will be distributed across multiple locations.

Consequently, most AIoT projects will require an Agile method that can be scaled beyond a single team and potentially a single location. In the discussion above, different approaches for scaled Agile development were introduced, including Scrum of Scrums (SoS), Large Scale Scrum (LeSS) and the Scaled Agile Framework (SAFe). We will discuss later how they can be adapted to overcome some of the other limitations of AIoT development.

Impediment 3: Sourcing & External Dependencies

Many AIoT-enabled products have a complex supply chain, often using different suppliers for cloud software, edge software, embedded hardware and software, AI, and telecommunications infrastructure. The combination of technical complexity and supply chain complexity can be difficult to manage in a purely agile setting. Sourcing can already be a complex topic in normal software projects. With the added complexities of an AIoT project, any form of Agile organization will have to be closely aligned with the sourcing process.

In some organizations, Agile teams will have a high level of autonomy over the sourcing process. In a matrix-type organization as described above, a central sourcing team will have a high level of influence and potentially little experience with this type of project. If the AIoT product is a large-scale, strategic product, it is very likely that procurement will dominate the first one or even two years of development. In addition, lead times for hardware acquisition can be much longer than those of software components.

Most likely, the greatest challenge from an Agile perspective is that most sourcing projects require extensive documentation of requirements or even solution designs. Many procurement organizations still see fixed-price offers as the preferred way of sourcing complex solutions. The level of documentation detail needed for most fixed-price contracts runs directly diametric to the low-documentation mantra of the Agile world.

Consequently, the product team and the procurement team will have to find ways to satisfy both of their needs, e.g., through creative solutions such as "fixed-price agile contracts" (e.g., a time-and-material contract with a performance-based compensation component based on individual sprint performance).

Finally, another major issue is SLAs (Service Level Agreements) and warranties. The buyer typically wants this fixed as early in the development cycle as possible (ideally as part of the contract), while the supplier will have a hard time agreeing to any SLA or warranty if the requirements are not fixed.

Impediment 4: Artificial Intelligence

While AI is quickly becoming a popular element of modern product development, there are currently no well-established methodologies for AI development. Some teams are applying methods such as CRISP-DM, KDD, or CPMAI to AI projects. However, there is not yet a well-documented and established method available that explains how to combine agile and AI. Which is not to say that it cannot be done, but the lack of proven agile methods for AI development is certainly an inhibitor for AIoT & Agile.

Impediment 5: Hardware/Embedded

Most AIoT projects require a combination of traditional (back-end) software development with some form of edge or embedded software development. In many cases, custom hardware development and manufacturing will also be a requirement.

If the edge platform is a powerful, server-like system, e.g. running Linux and C++ or even high-level languages, the difference from an Internet or enterprise software project might not be as high. However, in many cases, AIoT projects also require custom embedded software and hardware, which is a different animal all together, and many Agile principles cannot be directly applied. Embedded software is typically very technical, e.g., focusing on hardware-specific drivers or controllers, running on specialized real-time operating systems. In Addition, hardware development follows its own rules. Some differences include:

- Software in general is easier to change than hardware designs
- It is not possible to change hardware after it has been manufactured
- Higher-level hardware designs must often incorporate existing, standardized lower-level parts
- The lead times, acquisition times and feedback loops for hardware components are much longer
- Time to first full integration test with custom hardware usually significantly longer than with software

All of the above does not mean that an Agile mindset cannot be applied to embedded hardware/software development. However, an Agile embedded approach will have to take these specifics into consideration and adapt the methodology accordingly.

Impediment 6: First Time Right

Most AIoT-enabled products have some components and/or features which have to be "First Time Right", e.g., because they will be deployed on assets in the field and cannot be easily changed afterwards. This can make it very difficult to apply agile principles: in the magic project triangle, the agile approach usually aims to fix the time and budget side of things, while the features/scope are seen as variables. For First Time Right features, this approach does not work.

Impediment 7: Functional Safety

Finally, if the AIoT-enabled product has Functional Safety requirements, this can impose another significant impediment on agility. Many Functional Safety standards are based on a "requirements first, code later" philosophy, while Agile is focused on continuous delivery of working software, but not on extensive documentation and detailed requirements management. The Agile V-Model introduced by the AIoT Framework addresses this by providing a bridge between the two worlds.

23.1.2   Conclusions

As seen from this discussion, adopting an end-to-end agile approach in an AIoT project or AIoT product development organization will not be straightforward. The question is - does it make sense at all? An interesting tool to get some answers to this question is the Stacey Matrix [21], which is designed to help understand and manage the factors that drive complexity in a project (Fig. 23.4).

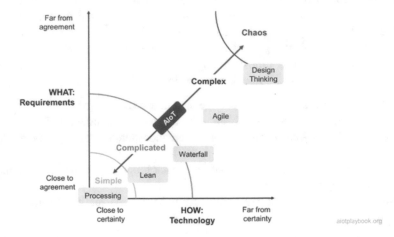

Fig. 23.4 Stacey matrix

The Stacey matrix is based on two dimensions: uncertainty of requirements (WHAT) and uncertainty of technology (HOW). The matrix has different areas, from "simple" (high level of certainty in both dimensions) to "anarchy" (far from certain). The matrix is often used to position different methods to these areas. For example, simple tasks can be addressed by a basic processing approach. Lean methods can help optimize well-understood tasks and eliminate waste in these standardized processes. Standard project methods can be utilized up to a certain amount of complexity. After this, agile is recommended. For near-chaotic or early-stage project phases, design thinking is recommended. This might be slightly black-and-white, but you get the picture.

However, the point with many AIoT initiatives is that they will have tasks or subprojects across the entire Stacey matrix. Take, for example, the manual labeling of one million images as input for an ML algorithm. This is a highly repetitive task with low tech and clear requirements. It seems unlikely that this task will benefit hugely from an agile approach. Or take the retrofit rollout of sensor packs to 10,000 assets in the field. Again, something with clear technical and functional requirements. On the other hand, take a "We will use AI to address xyz" statement from a management pitch. This will be in the upper right corner and again require a completely different approach.

Based on his experience as Project Management Process Owner at Bosch, Stephan Wohlfahrt has come to the following conclusions: *Use Agile wisely, not only widely! Always be aware that a plan-driven or hybrid approach might be the better fit for a particular project phase or sub project.*

An organization that has a lot of experience with both traditional project management and agile methods is the Project Management Institute (PMI). To address the challenges described above, the PMI has developed Disciplined Agile (DA) [22], which is a hybrid tool kit that harnesses Agile, Lean, and traditional strategies to provide the best way of working (WoW) for a team or organization. DA aims to be context-sensitive: rather than prescribing a collection of "best practices", its goal is to teach how to choose and later evolve a fit-for-purpose "WoW", which is the best fit in a given situation.

Scott Ambler is the Chief Scientist of Disciplined Agile at Project Management Institute. His advice on AIoT is as follows: *To succeed with AIoT, or any complicated endeavor, your team requires a fit-for-purpose approach. AIoT teams address complex problems, often ones where life-critical regulations apply and where your solution involves both hardware and software development, so you must be pragmatic in how you choose your WoW. You want to be as effective as you can be, and to do that you must choose an appropriate life cycle and practices that support it. AIoT teams face a competitive market and a rapidly evolving environment, so you must improve continuously via an experimentation-based, validated-learning strategy. Unlike agile frameworks that prescribe a collection of "best practices" DA instead teaches you how to choose the practices that are best for you – DA helps you to get better at getting better. You want your teams to work in a manner that is best for them, which implies that all of your teams will work differently and will constantly evolve how they work.*

The following introduces the Agile V-Model, which is a combination of agile practices and the V-Model, designed to specifically address many of the challenges discussed before. Since the Agile V-Model focuses on the project/product level, it can be combined with enterprise practices such as DA.

23.2 Agile AIoT Organization (Fig. 23.5)

The AIoT product organization must combine scaled agile methods with those methods that are better suited to address the aforementioned impediments to agility. As outlined in the book *The Connected Company* [23], companies which successfully address the fast speed of technical innovation must be more like complex, dynamic systems that can learn and adapt over time. The following provides an overview of how the *Digital Playbook* suggests addressing agility in the context of AIoT.

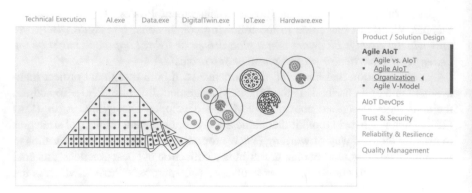

Fig. 23.5 Agile AIoT organization

23.2.1 *Industry Best Practices*

Many of the Cloud hyper-scalers have adopted best practices for customer-centric, agile product innovation. There are many examples from companies like Apple, Google, and Amazon. Some of the core elements of these best practices include a culture which is open for innovation, agile organizational structures and effective support mechanisms.

The organizational culture of these organizations is often described as being centered around a strong - if not obsessive - customer focus. Empowerment of employees with a strong "builder" mentality also plays an important role. Another point is calculated risk taking, which often goes hand-in-hand with a "fail early, fail often" mentality. Especially the latter is a point which is often very difficult to handle for companies with an industrial or manufacturing focus.

Another point are organizational structures which support agility. Again, often a weak point of large, industrial incumbents. Companies like Amazon are promoting small, empowered teams. The term „2 Pizza Team" was coined by Jeff Bezos himself - referring to the fact that a team should not include more people than can be fed by 2 pizzas. Another important point is single threaded ownership. This means that leaders should be able to focus 100% on a single deliverable - without having to worry about dependencies to other parts or the organization. This comes along with decentral decision making: Avoiding large decision-making bodies as they are known from many industrial incumbents, and instead empowering the teams and product owners.

These agile organizational structures must be supported by effective mechanisms. Again, looking at an Amazon example, which is the "Working backwards" process. In this process, the team starts with creating a press release which describes the outcome from the end customer's perspective, including information about the problem, how current solutions are failing, and why the new product will solve this problem. Other methods include Design Thinking, Lean UX and Customer Journey Mapping.

Finally, the agile organization must be supported by an architecture which is flexible and scalable to support growth and change. For many cloud organizations, this means a clear focus on microservices with well-defined APIs as the only means for access. In addition, these services must be made available as self-service platforms, to ensure scalability without depending on human administrators or gatekeepers. For many organizations which are not cloud-native, this is also an issue. Furthermore, combining hardware with software in AIoT-enabled products or solutions can also make it very hard to find flexible and easily changeable architectures (Fig. 23.6).

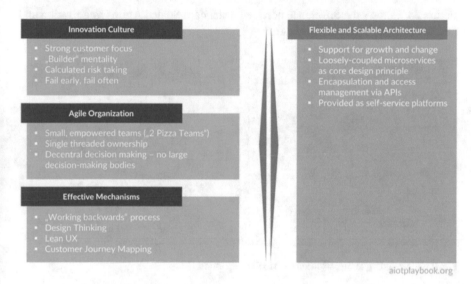

Fig. 23.6 Industry best practices

23.2.2 Scaled Agile Organizations and AIoT

There are a number of different agile frameworks designed to address the issue of scaling agile to a level where it can support larger organizations and multiple teams with complex dependencies. Examples include SAFe (the Scaled Agile Framework) and Less (Large Scale Scrum). The basic idea of most of these scaled agile approaches is to establish a leadership team that usually consists of a lead architect, a product manager (or chief product owner), and an engineering lead. Together, they help orchestrate the work by agile teams, which usually work relatively independently within a loosely coupled, agile organization.

This approach makes sense in general and is followed by the AIoT framework. To address the specifics of AIoT, two additions are made:

- A dedicated product coordinator is added to the leadership team. This role will be responsible for ensuring that all dependencies are properly managed, both internally and externally.
- Two types of agile teams are introduced: Feature Teams and AIoT Technical Workstreams. The Feature Teams take end-to-end responsibility for functional features, and the Technical Workstreams provide the foundations. Sometimes the Technical Workstreams are the home for experts with different technical skills (cloud, edge, mobile apps, etc.), which are then assigned to different Feature Teams, depending on the demand.

Figure 23.7 shows the difference between a standard Scaled Agile Organization and the Agile AIoT Product Organization.

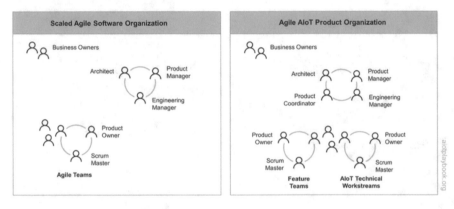

Fig. 23.7 Scaled agile organization vs AIoT organization

23.2.3 Feature Teams Versus Technical Workstreams

An AIoT product usually requires a technical infrastructure that includes cloud and edge capabilities, IoT communication services, and an AIoT DevOps infrastructure. In addition, many cross-cutting topics, such as end-to-end security, OTA (Over-the-Air Updates), etc. must be addressed. This is what the AIoT Technical Workstreams will need to provide and maintain.

The Feature Teams are then responsible for building end-to-end functionality based on the technical infrastructure. For this to occur in a coordinated way, the product manager must work closely with the product owners of the different feature teams (to map the end-to-end epics from the story maps to user stories that are managed in the backlogs of the different feature teams). The lead architect must ensure that the different feature teams contribute to a consistent end-to-end architecture. The product coordinator, the engineering lead, and the different product owners should collaborate to manage key dependencies and meet global milestones (Fig. 23.8).

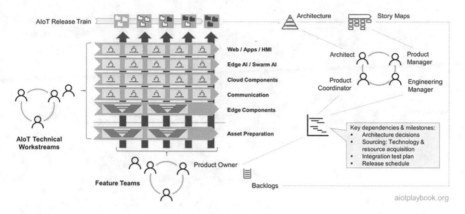

Fig. 23.8 AIoT feature teams vs. technical workstreams

23.2.4 Minimum Viable Teams

When setting up a new team — whether a feature team or a technical workstream — a key question is how to staff it. This is why the concept of the Minimum Viable Team (MVT) is so important. The purpose of the MVT is to make the initial team as lean as possible, and allow the core team to pull in additional resources when needed.

Always keep in mind Conway's Law, which describes the phenomenon where the organizational structure becomes the blueprint for the architectural structure. For example, if you have a database expert on the team the final design will probably include a database, regardless of whether one is needed or not. As soon as organizations decide who will be on the team, they are in effect designing the system.

This can be addressed by the concept of the MVT. The only caveat is that (especially in larger organizations) it can be difficult to obtain additional resources on demand. This is the reason many team leads have a tendency to acquire resources when the opportunity arises.

23.2.5 Leadership Roles

The following describes the key leadership roles in the Agile AIoT Product Organization: AIoT Product Manager, AIoT Product Engineering Lead, AIoT Product Architect, and AIoT Product Coordinator.

AIoT Product Manager
Product Management for smart, connected products can build on existing good practices, which are summarized in Fig. 23.9.

Fig. 23.9 Product management

AIoT Product Engineering Lead

The Product Engineering Lead is effectively responsible for ensuring that the different AIoT product teams are together continuously delivering integrated product increments. Depending on the chosen setup — loosely coupled teams or a more hierarchical organization — they will directly or indirectly coordinate and orchestrate the delivery of the product increments.

Some of the key tasks include:

- Management of end-to-end product engineering roadmap
- Alignment of product vision with product roadmap and backlogs
- Facilitation of cross-team planning events
- Oversee continuous delivery pipeline and efficient AIoT DevOps - covering edge and cloud, as well as code and AI models
- Coordination/support of technology and resource acquisition
- Escalation and tracking of road-blockers
- Ensure UX principles are followed
- Ensure establishment of Quality Management - including QM for the AI-elements of the system
- Creation and tracking of key engineering metrics - including metrics for the AI-part of the system
- Coaching and guiding the engineering staff - ensuring that hardware development, traditional software development and AI-development are working hand in hand
- Collaboration with a product coordinator on dependency management, risk management, and cost management

AIoT Product Architect

The AIoT Product Architect must define and maintain the end-to-end architecture and design of the product. They must work closely with the different AIoT development teams and other key stakeholders, focusing only on architectural decisions that are of relevance from a cross-team perspective. Architectural decisions that have no impact outside of an individual team should be made by the team itself.

It is important to note that the AIoT Framework does not propose an excessive, RUP/waterfall-style model depth, as can be seen when looking at the individual templates. The general scheme of collaboration between the different project stakeholders in the architecture management process is shown in Fig. 23.10.

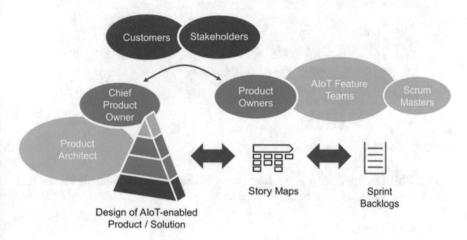

Fig. 23.10 AIoT solution architecture process

The key to success is to keep the AIoT Solution Architecture on a level of detail where it is meaningful, but not overly complex. The agile teams must be able to apply their own mechanism (e.g., demand-driven design spikes) to derive requirements for their backlog and provide feedback to the overarching architecture in return.

AIoT Product Coordinator
To support the Product Manager, Engineering Lead and Product Architect in their work, AIoT recommends installing an overall Product Coordinator in a back office management role. The key tasks of this coordinator are summarized in Fig. 23.11.

Fig. 23.11 AIoT project management

PMI PMBOK provides a good description of the different knowledge areas that a product coordinator must be able to support.

23.3   Agile V-Model (Fig. 23.12)

The V-Model is a software development method often found in areas with high requirements on safety and security, which are common in highly regulated areas. Combining the traditional V-Model with a disciplined agile approach promises to allow as much agility as possible, while addressing the issues often found in AIoT initiatives: complex dependencies, different speeds of development, and the "first time right" requirements of those parts of the system which cannot be updated after the Start of Production (SOP).

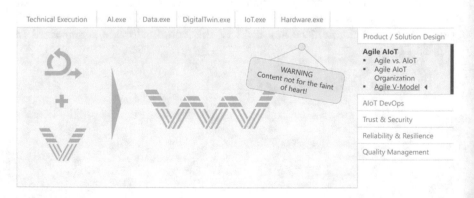

Fig. 23.12   1.0 Agile V-Model.png

23.3.1   Recap: The V-Model

The V-model is a systems development lifecycle which has verification and validation "built in". It is often used for the development of mission critical systems, e.g.,
in automotive, aviation, energy and military applications. It also tends to be used in
hardware-centric domains. Not surprisingly, the V-model uses a v-shaped visual
representation, where the left side of the "V" represents the decomposition of
requirements, as well as the creation of system specifications ("definition and
decomposition"). The right side of the "V" represents the integration and testing of
components. Moving up on the right side, testing usually starts with the basic verification (e.g., unit tests, then integration tests), followed by validation (e.g., user
acceptance tests) (Fig. 23.13).

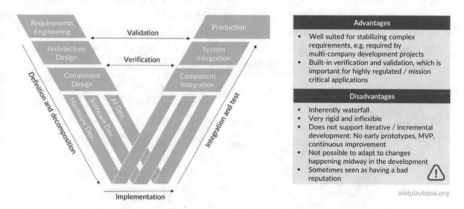

Fig. 23.13 V-Model

When applying the V-model to AIoT, it needs to take different dimensions into
consideration; usually including hardware, software, AI, and networking. In addition
to the standard verification tests (unit tests, integration tests) and validation tests (user
acceptance and usability tests), the V-model for AIoT also needs to address interoperability testing, performance testing, scalability testing and reliability testing. The
highly distributed nature of AIoT systems will pose specific challenges here.

Test automation is key to ensure a high level of test efficiency and test coverage.
On the software side, there are many tools and techniques available to support this.
In the AI-world, these kinds of tools and techniques are only just beginning to
emerge, which makes it likely that a more custom approach will be required. In the
embedded and hardware world, simulation techniques such as Hardware-in-the-
Loop (HIL), Software-in-the-Loop (SIL) and Model-in-the-Loop (MIL) are well
established. However, most AIoT products will also require testing of the actual
physical product and how well they perform in the field in different types of environments. Again, this will be a challenge, and some ingenuity will be required to
automate testing of physical products wherever possible.

23.3.2 Evolution: The Agile V-Model

The AIoT framework aims to strike a good balance between the agile software world and the less agile world of often safety-critical, complex and large(r)-scale AIoT product development with hardware and potentially manufacturing elements. Therefore, it is important to understand how an agile method works well together with a V + V-centric approach such as the V-model. The logical consequence is the *Agile V-model*. Combining agile development with the V-model is not a contradiction. They can both work very well together, as shown in the figure following:

- Agile methods use story maps including epics, themes, features and user stories for logical decomposition. This maps well to the left side of the V
- Continuous Integration / Continuous Test / Continuous Delivery are inherently agile methods, that map well to the right side of the V
- The key assumption is that the V-model is not used like one large waterfall approach. Instead, the Agile V-model must ensure that the sprints themselves will become Vs according to the V-model (Fig. 23.14)

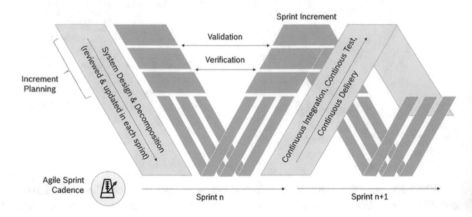

Fig. 23.14 Agile V Model

There are two options to implement the latter:

- Each sprint becomes a complete V, including development and integration/test
- The agile schedule introduces the concept of dedicated integration sprints.
- One V becomes 2 sprints: one development sprint, one integration sprint
- There are pros and cons to both approaches
- The complexity and scale of the project will surely play a role in determining the best setup

For most projects / product teams, it is recommended that development and integration are combined in a single sprint ("v-sprint"). Only for projects with a very high level of complexity and dependencies, e.g., between components developed by different organizations, is it recommended to alternate between development and integration sprints. The latter approach is likely to add inefficiencies to the development process, but could be the only approach to effectively deal with alignment across organizational boundaries.

23.3.3   The ACME:Vac Vacuum Robot Example

To illustrate the use of the Agile V-Model, the realistic yet fictitious ACME:Vac example is introduced. This is a robot vacuum cleaning system that combines a smart, connected vacuum robot with a cloud-based back end, as well as a smart app for control.

Modern robot vacuum cleaners are very intelligent, connected products. Even the most basic versions provide collision, wheel, brush and cliff sensors. More advanced versions use visual sensors combined with a VSLAM algorithm (Visual Simultaneous Location and Mapping). The optical system can identify landmarks on the ceiling, as well as judge the distance between walls. The most advanced systems utilize LIDAR technology (Light Detection and Ranging) to map out their environment, identify room layouts and obstacles, and serve as input for computing efficient routes and cleaning methods. For example, the robot can decide to make a detour vs. switching into the built-in "climb over obstacle"-mode. Another example is the automatic activation of a "carpet boost" mode. IoT-connectivity to the cloud enables integration with user interface technology such as smart mobile devices or smart home appliances for voice control ("clean under the dining room table"). Edge AI algorithms deployed on the robot are used to control these processes in advanced models. A complete design of the ACME:Vac is provided in the Product / Solution Design section (Fig. 23.15).

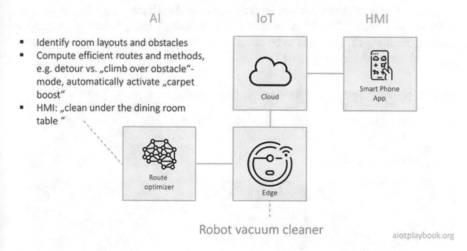

Fig. 23.15   Example: robot vacuum cleaner

23.3.4 Applying the Agile V-Model to ACME:Vac

The following provides a description of how the ACME:Vac example is developed using the Agile V-Model. This discussion will start with a look at a "Sprint 0", where the foundations are laid. Then a "Sprint n" is discussed in detail, with an outlook on the transition to the next sprint ("Sprint n+").

Sprint 0

Many scrum masters are starting their projects with a "Sprint 0" as a preparation sprint to initiate the project (ignoring the quasi-religious discussion whether this is a good practice or a "cardinal sin" for the moment, since we have to start somewhere...). In the case of ACME:Vac, two working results arise: the initial story map and the initial component architecture. These will be important elements for the planning of the subsequent sprints (Fig. 23.16).

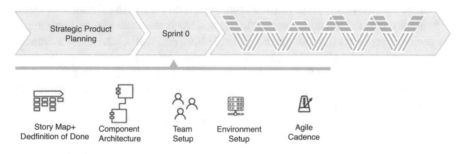

Fig. 23.16 Sprint Zero

Initial Story Map

According to the story map structure proposed by the AIoT Framework, the story map for the vacuum robot system includes epics and features on the top level. The epics include Human/Machine Interfaces, the actual cleaning functions, management of the maps for the areas to be cleaned, navigation/sensing, system configuration, and status/history (Fig. 23.17).

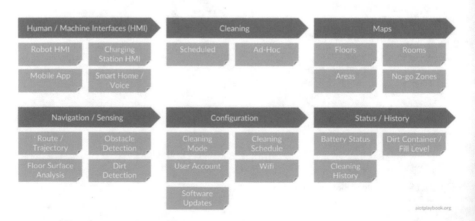

Fig. 23.17 Example: initial story map

Initial Component Architecture
The Component Architecture for the ACME:Vac highlights the key functional components in three clusters: the robot itself (Edge), the cloud back end, and the smartphone app. On the robot, two embedded components provide control over the robot functions, as well as access to the sensor data. Higher-level components include the control of the robot movements (based on AI/ML, potentially with a dedicated hardware), the robot configuration, as well as remote access to the robot APIs. The cloud services include basic services for managing map data, status/history data, as well as user and robot configuration data. The robot control component enables remote access to the robot. Finally, the smart phone/mobile app provides components for robot configuration and map management, all accessible via the main screen of the app (Fig. 23.18).

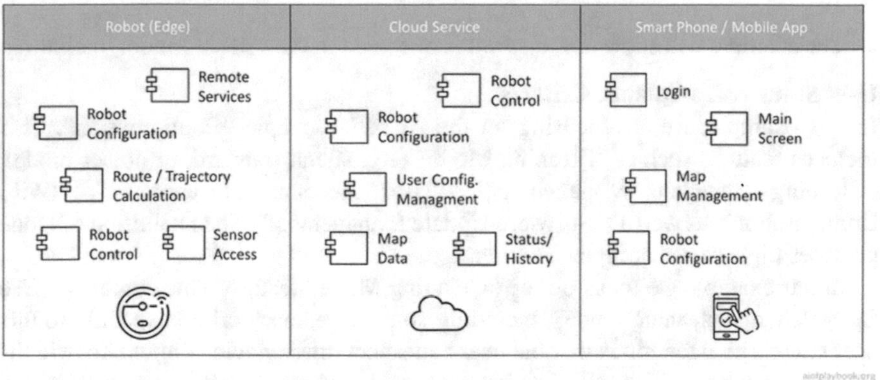

Fig. 23.18 Example: initial component architecture

Note that — in line with the agile "working code over documentation" philosophy — the component architecture documentation does not need to be very detailed. Only the main functional components are listed here. Depending on the project, usage dependencies can be added if such levels of detail are deemed relevant and the maintenance of the information is realistic.

Sprint n
Next, we want to jump right into the middle of the ACME:Vac development by fast-forwarding to a "Sprint n", which will focus on a single feature to illustrate the approach in general (Fig. 23.19).

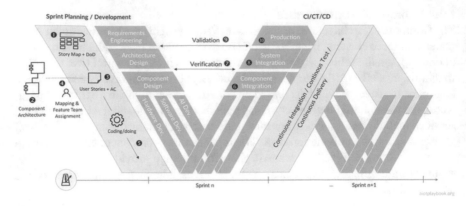

Fig. 23.19 Agile V-Model

The example will show how a sprint is executed in the agile V-Model, including:

User Story & Acceptance Criteria
In this example, we are focusing on the ACME:Vac Epic "Configuration". This includes features such as "Cleaning Mode" (e.g. silent, standard, or power mode), "Cleaning Schedule Management", "User Account Management", "WiFi Configuration", as well as "Software Update Management". The Definition of Done provides higher level acceptance criteria.

In our example we focus on the "Cleaning Mode" feature. This contains a Use Story "change cleaning mode", including some acceptance criteria specific to this user story. The intention is that the user can select different cleaning modes via his smart phone app, which will then be supported by both the ad-hoc as well as the scheduled cleaning modes (Fig. 23.20).

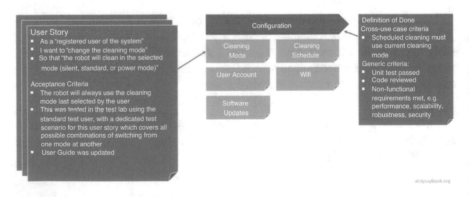

Fig. 23.20 Example user story

Mapping User Story to Components and Feature Team

Having defined the user story, the next step is to identify which components are required for the story. The "Change Cleaning Mode" story will require a robot configuration component on the smartphone. This will need to interact with the robot configuration component in the cloud. In order to record the change, an interaction with the status / history component is required. Finally, the remote service component on the robot will receive the selected mode, and configure the robot accordingly (Fig. 23.21).

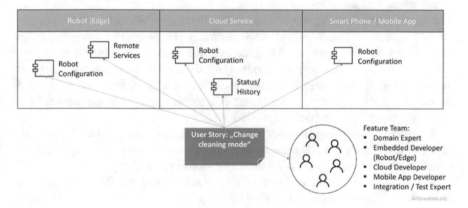

Fig. 23.21 Mapping user story to components and feature team

Based on the analysis of the functional components, the required members of the feature team for this user story can be identified. They include a domain expert, an embedded developer for the components on the robot, a cloud developer, a mobile app developer, and an integration/test expert. Note that some organizations strive to employ full-stack developers. However, in this case it seems unlikely that a single developer will have all the required skills.

Implementation and CI/CT/CD

Implementation in an AIoT initiative can include many components, including software development, AI/ML development, data management, HW design and engineering, or even manufacturing setup. The various tasks often have to be executed at different development speeds, e.g., because hardware is usually not evolving as fast as software. Because of this, it might not always be possible to create a potentially shippable product increment. The Agile V-Model recommends that if this is a case, at least mockup implementations of the public interfaces should be provided for integration testing.

Each sprint needs to fully support Continuous Integration, Continuous Testing and Continuous Deployment (CI/CT/CD), with the limitations just discussed in mind. A key element of CI is component integration, which usually should be done with a focus on different user stories, i.e., it should focus on integrating and testing the set of components required for a particular user story. This often means that these components are embedded into an environment which simulates auxiliary components and/or production data. This should be handled automatically by the CI (Continuous Integration) infrastructure.

If the component integration including tests was successful, the components can be integrated into the entire system. This means integrating all components which are changed or created by the current sprint need to be integrated in order to create the next, potentially shippable increment of the system.

Verification & Validation

For some people in the agile community, Verification & Validation (V&V) is considered to be an outdated concept, while for many people with an enterprise and/or functional safety background, it is good practice. The Agile V-Model aims for a pragmatic view on this. Componentization (see the Divide & Conquer section) must support an approach where functional components with different levels of QM/ functional safety requirements are separated accordingly, so that the most suitable V&V approaches can be applied individually.

Traditionally, validation is supposed to answer the question *"Are we building the right product?"*, while verification focuses on *"Are we building the product right?"*, even though the boundary between these questions can be blurry.

In the Agile V-Model, verification differentiates between the micro and the macro-level. The micro-level is defined by the Acceptance Criteria of the individual user stories, while the macro-level is focusing on the Definition of Done, which usually applies across individual user stories. Validation, on the other hand, focuses on user acceptance. In AIoT, this can involve quite laborious lab and field tests (Fig. 23.22).

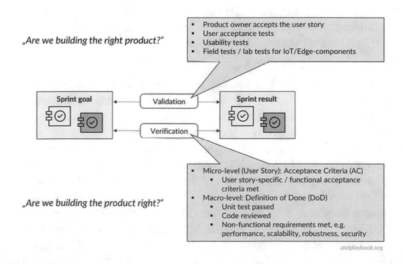

Fig. 23.22 V&V details – explanation

Sprint n + 1
At the end of a sprint, the sprint retrospective should summarize any key findings from the sprint's V&V tasks that need to be carried over to the next sprint. For example, findings during User Acceptance Tests might require a change to a user story. Or a feature team might have decided on an ad hoc change to the planned component architecture. These kinds of findings must either be reflected by changes to the backlog (for user story definitions), or by updating the architecture plans and documentation.

Summary
In summary, the Agile V-Model is designed to support the adoption of agile practices in an AIoT environment, where we usually find some significant inhibitors of a "pure" agile approach. Each v-sprint combines a normal, agile sprint with a sprint-specific planning element. This allows us to address the complexities inherent to most AIoT initiatives, as well as the typical V&V requirements often found in AIoT.

At the end of each sprint, the sprint results must be compared to the original sprint goals. Story maps and component landscapes must be adapted according to the results of this process. The updated version then serves as input to the planning of the next sprint (Fig. 23.23).

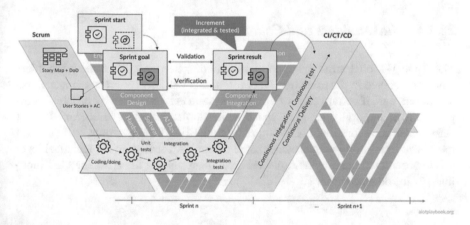

Fig. 23.23   Verification and validation – details

23.3.5   Decoupling Development

To master the complexity of an AIoT product or solution, it is important to apply an effective "Divide & Conquer" strategy. This requires decoupling both on the development and on the technical level. In order to decouple the development, an interface-first strategy must be implemented. This will allow for the development and testing of components independently of each other, until they have reached a level of maturity so that they can be fully integrated and tested.

This will also be especially important in AIoT because it will not always be possible to test all components in the context of the full system. This is why many AIoT developments utilize different test simulation approaches. These can range from simple interface simulations (e.g., via REST tools) to sophisticated virtual simulations (e.g., allowing us to test an algorithm for autonomous driving in a virtual world). Since field tests can be time-consuming and expensive, this is important in AIoT (Fig. 23.24).

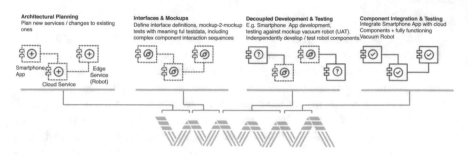

Fig. 23.24 Decoupling development

23.3.6 Stakeholders and Collaboration

The Agile V-Model aims to support collaboration between different stakeholders who are involved in the execution. Very often, this will involve cross-company collaboration, e.g., if certain parts of the AIoT system are developed by different companies. It is important to think about which tools should be used for efficient stakeholder collaboration, both within individual stakeholder groups and between them. The figure following provides some examples. Finally, on the technical level it is important to note that the Agile V-Model must support the different AIoT pipelines, as introduced earlier (Fig. 23.25).

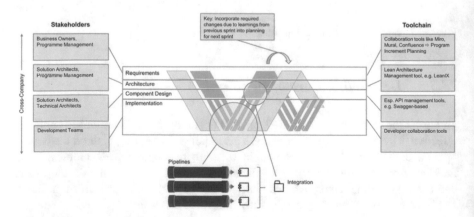

Fig. 23.25 Agile V-Model: stakeholders and toolchain

23.3.7   Agile V-Model and AIoT

Finally, this discussion should conclude by mentioning what is probably the biggest difference between a "normal" software project and an AIoT project: the Start of Production, or SOP. This is the point in time when mass production of the smart, connected products is starting, and they are leaving the factory to be deployed in the field. Alternatively, the start of the roll-out of retrofitting the AIoT solution to existing assets in the field. Any required changes to the hardware configuration of the product or solution will now be extremely difficult to achieve (usually involving costly product recalls, which nobody wants). In a world where manufacturers want to utilize the AIoT to constantly stay in contact with their products after they have reached the customer, and provide the customer with new digital services and applications — rolled out via OTA — it becomes extremely important to understand that the V&V process does not stop with the SOP, at least not for anything that is software. With software -defined vehicles, software-defined robots, and software-defined everything, this is quickly becoming the new normal (Fig. 23.26).

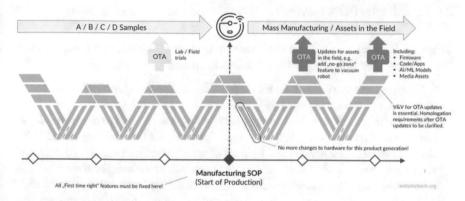

Fig. 23.26   Agile V-Model and SOP

23.3.8   Issues and Concerns

The proposed Agile V-Model approach has provoked intense discussions between advocates of the different worlds. Some of the arguments against the approach have been captured in Fig. 23.27. Some comments and counterarguments are included as well. We hope that the discussion will continue, and we will be able to capture more pros and cons (Fig. 23.27).

Issue / Warning about the V-Model approach	Remarks
Don't fall back into RUP-style, super detailed documentation which is costly and always outdated.	Very true, this must definitely be avoided. The examples provided here are hopeful showing how this can be avoided by keeping it on the right level of detail and combining it with agile best practices. Still, in complex projects a certain level of high-level architecture documentation will be required to align all stakeholders.
This is introducing too much processual overhead.	Maybe true. Depends on how you set up the Agile V-Model in your organization. Plus, some organizations still feel more comfortable with at least a certain level of process.
You should not centralize requirements management and architecture decisions. The outcome should be driven by the Definition of Done.	Partly true. The Agile V-Model is proposing to combine Definition of Done (per feature/use case) or Acceptance Criteria (per User Story) with a lightweight, overarching system design to ensure cross-team alignment.
Sprint planning must be done before the next sprint starts.	True. The planning of the next sprint must start while you move up the right side of the previous V.
Strict decomponentization too early in the development cycle can cause problems further downstream, if not done right.	True. Finding the right structure for the component landscape is key, and often requires multiple iterations. This is a strong point of the Agile V-Model, because the high-level component landscape is revisited at the beginning of each v-sprint.
V-Model has a bad reputation, the name is tainted	Maybe true. The name was chosen to provoke a discussion. Let's see where it goes and whether we have to reconsider this.

aiotplaybook.org

Fig. 23.27 Issues and concerns regarding the Agile V-Model

23.3.9 Expert Opinion

Sebastian Helbeck is a VP and Platform Owner Power Tools Drive Train at Bosch Power Tools. The following his summarizes his thoughts on this topic.

Dirk Slama: *Do you see the agile world (cloud, etc.) and the waterfall world (e.g., embedded) currently working well together? What are the issues?*

Sebastian Helbeck: *What I am seeing is that many projects are currently developed parallel in these two worlds. This is based on the fact that in many cases either the embedded or the non-embedded part already exists, and the other part needs to be developed. In the future this needs to be done more seamlessly.*

DS: *How much do these worlds need about each other, and how can we ensure this?*

SH: *Currently everybody is focusing on understanding and fulfilling their own contributions and the interfaces around. In the future, the interfaces will change and will be even more complex. Edge computing will be a key driver here. Consequently, we need to have better knowledge about our own contributions and how they relate to the other side. As an analogy, FPGA technology can be used, which has tremendously changed the interfaces between HW and embedded SW.*

DS: *Can the different methods coexist, or do we need a common approach?*

SH: *Currently, they coexist and it works, but it is not ideal. This is why as the next step, we need to bring the different approaches closer together due to the different maturities of the product parts. To unleash the full power of AIoT, we need to rethink this and have to adopt a more holistic approach.*

DS: *Which role can the Agile V-Model play here?*

SH: *It can potentially play an important role here. It will combine the two worlds with a holistic view to give all the included stakeholders a transparent overview of the complete system. For the future, we might need to think about a new name since the combination of the worlds will bring us to a new dimension of understanding of our AIoT systems.*

Chapter 24
Holistic AIoT DevOps

Dirk Slama

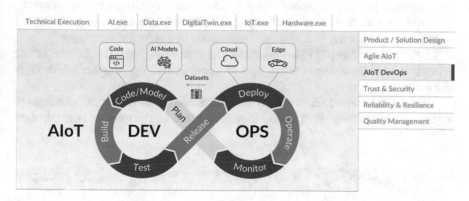

Fig. 24.1 AIoT - DevOps and infrastructure

The introduction of DevOps — together with Continuous Integration/Continuous Delivery (CI/CD) — has fundamentally changed the way software is developed, integrated, tested, and deployed. DevOps and CI/CD are key enablers of agile development. However, today's DevOps practices predominantly focus on cloud and enterprise application development. For successful AIoT products, DevOps will need to be extended to include AI and IoT (Fig. 24.1).

D. Slama (✉)
Ferdinand Steinbeis Institute, Berlin, Germany
e-mail: dirk.slama@bosch.com

24.1   Agile DevOps for Cloud and Enterprise Applications

DevOps organizations breakdown the traditional barriers between development and operations, focusing on cross-functional teams that support all aspects of development, testing, integration and deployment. Successful DevOps organizations avoid overspecialization and instead focus on cross-training and open communication between all DevOps stakeholders.

DevOps culture is usually closely aligned with agile culture; both are required for incremental and explorative development.

Continuous Integration / Continuous Delivery (CI/CD) emphasize automation tools that drive building and testing, ultimately enabling a highly efficient and agile software life cycle. The Continuous Integration (CI) process typically requires commitment of all code changes to a central code repository. Each new check-in triggers an automated process that rebuilds the system, automatically performs unit tests, and executes automated code-quality checks. The resulting software packages are deployed to a CI server, with optional notification of a repository manager.

Continuous Testing (CT) goes beyond simple unit tests, and utilizes complex test suites that combine different test scripts to simulate and test complex interactions and processes.

Finally, Continuous Delivery (CD) uses Infrastructure-as-Code (IaC) concepts to deploy updated software packages to the different test and production environments (Fig. 24.2).

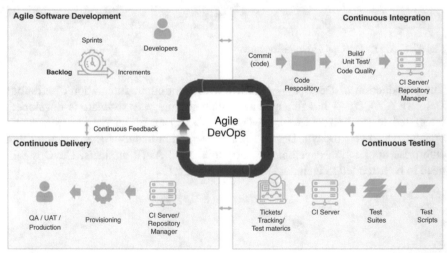

Fig. 24.2 Agile DevOps for cloud and enterprise applications

24.2   Agile DevOps for AI: MLOps

The introduction of AI to the traditional development process is adding many new concepts, which create challenges for DevOps:

- New roles: data scientist, AI engineer
- New artefacts (in addition to code): Data, Models
- New methods/processes: AI/data-centric, e.g., "Agile CRISP-DM", Cognitive Project Management for AI (CPMAI)
- New AI tools + infrastructure

The development of AI-based systems also introduces a number of new requirements from a DevOps perspective:

- Reproducibility of models: Creating reproducible models is a key prerequisite for a stable DevOps process
- Model validation: Validating models from a functional and business perspective is key
- Explainability (XAI, or 'explainable AI'): How to ensure that the results of the AI are comprehensible for humans?
- Testing and test automation: AI requires new methods and infrastructure
- Versioning: Models, code, data
- Lineage: Track evolution of models over time
- Security: Deliberately skewed models as new attack vector/adversarial attacks
- Monitoring and retraining: Model decay requires constant monitoring and retraining

Figure 24.3 provides an overview of how an AI-specific DevOps process can help in addressing many of the issues outlined above.

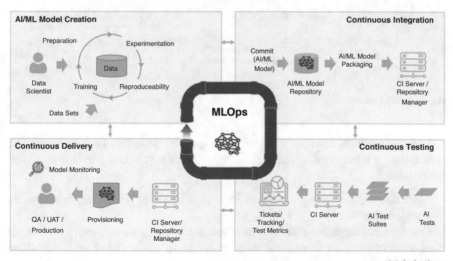

Fig. 24.3 AI DevOps

24.3   Agile DevOps for IoT

Finally, we need to look at the DevOps challenges from an IoT point of view. The main factors are:

- OTA: Over-the-Air updates (OTA) require a completely different infrastructure and process than traditional, cloud-based DevOps approaches
- Embedded Software & Hardware: The lifecycle of embedded hardware and software is very different from cloud-based software. Testing and test automation are possible, but require special efforts and techniques.

The OTA update process is described in more detail here. The figure following provides a high-level overview. The OTA Update process usually comprises three phases. During the authoring phase, new versions of the software (or AI models or other content) are created. The distribution phase is responsible for physical distribution (e.g., between different global regions) and the management of update campaigns. Finally, once arrived on the asset, the local distribution process ensures that the updates are securely deployed, and the updated system is validated (Fig. 24.4).

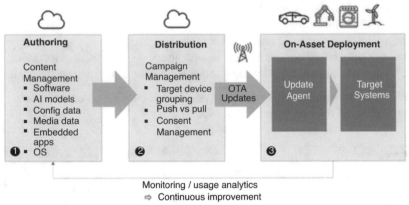

Fig. 24.4 OTA overview

Looking again at the four quadrant DevOps overview, this time from the IoT perspective, a number of differences compared to the standard DevOps approach can be seen:

- Agile development is structured to match the needs of an IoT organization, as discussed here
- Continuous Integration (CI) will usually have to cover a much more diverse set of development environments since it needs to cover cloud and embedded development

- Continuous Testing (CT) will have to address the test automation of embedded components, e.g. by utilizing different abstraction and simulation techniques such as HIL (hardware in the loop), SIL (software in the loop) and MIL (model in the loop)
- Continuous Delivery (CD) will have to utilize OTA not only for the production system but also for Quality Assurance and User Acceptance Tests

Finally, all of the above will also have to be examined from the perspective of Verification and Validation (Fig. 24.5).

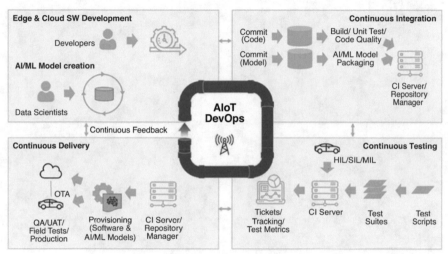

Fig. 24.5 AIoT DevOps

24.4 Agile DevOps for AIoT

The AIoT DevOps approach will need to combine all three perspectives outlined in the previous sections: Cloud DevOps, AI DevOps and IoT DevOps. Each of these three topics in itself is complex, and integrating the three into a single, homogeneous and highly automated DevOps approach will be one of the main challenges of each AIoT product. However, without succeeding in this effort, it will be nearly impossible to deliver an attractive and feature-rich product that can also evolve over time, as far as the limitations of hardware deployed in the field will allow. Utilizing OTA to evolve the software and AI deployed on the assets in the field will be a key success factor for smart, connected products in the future.

Chapter 25
Trust & Security

Dirk Slama

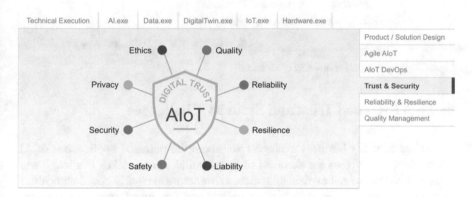

Fig. 25.1 Ignite AIoT - trust & security

Digital Trust (or trust in digital solutions) is a complex topic. When do users deem a digital product truly trustworthy? What if a physical product component is added, as in smart, connected products? While security is certainly a key enabler of Digital Trust, there are many other aspects that are important, including ethical considerations, data privacy, quality and robustness (including reliability and resilience). Since AIoT-enabled products can have a direct, physical impact on the well-being of people, safety also plays an important role (Fig. 25.1).

Safety is traditionally closely associated with Verification and Validation; which has its own, dedicated section in Ignite AIoT. The same holds true for robustness (see Reliability and Resilience). Since security is such a key enabler, it will have its own, dedicated discussion in this chapter, followed by a summary of AIoT Trust

D. Slama (✉)
Ferdinand Steinbeis Institute, Berlin, Germany
e-mail: dirk.slama@bosch.com

© The Author(s) 2023
D. Slama et al. (eds.), *The Digital Playbook*,
https://doi.org/10.1007/978-3-030-88221-1_25

Policy Management. Before delving into this, we first need to understand the AI and IoT-specific challenges from a security point of view.

25.1 Why Companies Invest in Cyber Security (Fig. 25.2)

Fig. 25.2 Why companies invest in cyber security

25.2 AI-Related Trust and Security Challenges

As excited as many business managers are about the potential applications of AI, many users and citizens are skeptical of its potential abuses. A key challenge with AI is that it is *per se* not explainable: there are no more explicitly coded algorithms, but rather "black box" models that are trained and fine-tuned over time with data from the outside, with no chance of tracing and "debugging" them the traditional way at runtime. While Explainable AI is trying to resolve this challenge, there are no satisfactory solutions available.

One key challenge with AI is bias: while the AI model might be statistically correct, it is being fed training data that include a bias, which will result in (usually unwanted) behaviour. For example, an AI-based HR solution for the evaluation of job applicants that is trained on biased data will result in biased recommendations.

While bias is often introduced unintentionally, there are also many potential ways to intentionally attack an AI-based system. A recent report from the Belfer Center describes two main classes of AI attacks: Input Attacks and Poisoning Attacks.

Input attacks: These kinds of attacks are possible because an AI model never covers 100% of all possible inputs. Instead, statistical assumptions are made, and mathematical functions are developed to allow creation of an abstract model of the real world derived from the training data. So-called adversarial attacks try to exploit this by manipulating input data in a way that confuses the AI model. For example, a small sticker added to a stop sign can confuse an autonomous vehicle and make it think that it is actually seeing a green light.

Poisoning attacks: This type of attack aims at corrupting the model itself, typically during the training process. For example, malicious training data could be inserted to install some kind of backdoor in the model. This could, for example, be used to bypass a building security system or confuse a military drone.

25.3   IoT-Related Trust and Security Challenges

Since the IoT deals with the integration of physical products, one has to look beyond the cloud and enterprise perspective, including networks and physical assets in the field. If a smart connected product is suddenly no longer working because of technical problems, users will lose trust and wish back the dumb, non-IoT version of it. If hackers use an IoT-connected toy to invade a family's privacy sphere, this is a violation of trust beyond the normal hacked internet account. Consequently, addressing security and trust for any IoT-based product is key.

The OWASP (The Open Web Application Security Project, a nonprofit foundation) project has published the OWASP IoT Top 10, a list of the top security concerns that each IoT product must address:

- Weak Guessable, or Hardcoded Passwords
- Insecure Network Services
- Insecure Ecosystem Interfaces (Web, backend APIs, Cloud, and mobile interfaces)
- Lack of Secure Update Mechanism (Secure OTA)
- Use of Insecure or Outdated Components
- Insufficient Privacy Protection
- Insecure Data Transfer and Storage
- Lack of Device Management
- Insecure Default Settings
- Lack of Physical Hardening

Understanding these additional challenges is key. However, to address them — together with the previously discussed AI-related challenges — a pragmatic approach is required that fits directly with the product team's DevOps approach. The result is sometimes also referred to as DevSecOps, which will be introduced in the following.

25.4   DevSecOps for AIoT

DevSecOps augments the DevOps approach, integrating security practices into all elements of the DevOps cycle. While traditionally many security teams are centralized, in the DevSecOps approach it is assumed that security is actually delivered by the DevOps team and processes. This starts with Security-by-Design, but also includes integration, testing and delivery. From an AIoT perspective, the key is to ensure that DevSecOps addresses all challenges presented by the different aspects of AIoT: AI, cloud/enterprise, network, and IoT devices/assets. Figure 25.3 provides an overview of the proposed AIoT DevSecOps model for AIoT.

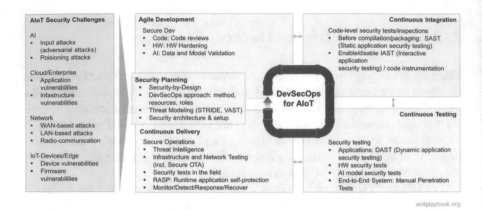

Fig. 25.3 DevSecOps for AIoT

DevSecOps needs to address each of the four DevOps quadrants. In addition, Security Planning was added as a fifth quadrant. The following will look at each of these five quadrants in detail.

25.5 Security Planning for AIoT

Security Planning for AIoT must first determine the general approach. Next, Threat Modeling will provide insights into key threats and mitigation strategies. Finally, the security architecture and setup must be determined. Of course, this is an iterative approach, which requires continuous evaluation and refinement.

25.5.1 DevSecOps Approach

The first step toward enabling DevSecOps for an AIoT product organization is to ensure that key stakeholders agree on the security method used and how to integrate it with the planned DevOps setup. In addition, clarity must be reached on resources and roles:

- Is there a dedicated budget for DevSecOps (training, consulting, tools, infrastructure, certification)?

- Will there be a dedicated person (or even team) with their security hat on?
- How much time is each developer expected to spend on security?
- Will the project be able to afford dedicated DevSecOps training for the development teams?
- Will there be a dedicated security testing team?
- Will there be external support, e.g., an external company performing the penetration tests?
- How will security-related reporting be set up during development and operations?

25.6 Threat Modeling

Threat Modeling is a widely established approach for identifying and predicting security threats (using the attacker's point of view) and protecting IT assets by building a defense strategy that prepares the appropriate mitigation strategies. Threat models provide a comprehensive view of an organization's full attack surface and help to make decisions on how to prioritize security-related investments.

There are a number of established threat modeling techniques available, including STRIDE and VAST. Figure 25.4 describes the overall threat modeling process.

First, the so-called Target of Evaluation (ToE) must be defined, including secu-

Fig. 25.4 Threat modeling

rity objectives and requirements, as well as a definition of assets in scope.

Second, the Threats & Attack Surfaces must be identified. For this, the STRIDE model can be used as a starting point. STRIDE provides a common set of threats, as defined in fig. 25.5 (including AIoT-specific examples).

	Threat	Property Violated	Threat Definition	AIoT Example
S	Spoofing identity	Authentication	Pretending to be something or someone other than yourself	Take over control of asset (e.g. car) by assuming identity of authorized user
T	Tampering with data	Integrity	Modifying something on disk, network, memory, or elsewhere	Reduce asset usage cost by falisifying asset usage records
R	Repudiation	Non-Repudiation	Claiming that you didn't do something or were not reponsible	Avoid taking responsibility in case of accident with asset
I	Information disclosure	Confidentiality	Providing information to someone not authorized to access it	Access detailed manufacturing performance records from competing manufacturer
D	Denial of service	Availability	Exhausting resources needed to provide service	Ground fleet of mobile assets (e.g. fleet of cars from mobility service provider)
E	Elevation of privilege	Authorization	Allowing someone to do something they are not authorized to do	Stop elevator on protected VIP floor

Fig. 25.5 STRIDE

The STRIDE threat categories can be used to perform an in-depth analysis of the attack surface. For this purpose, threat modeling usually uses component diagrams of the target system and applies the threat categories to it. An example is shown in fig. 25.6.

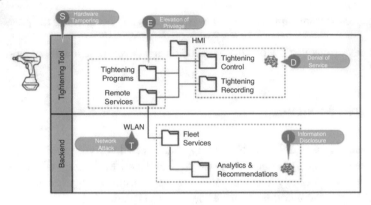

Fig. 25.6 Analyzing the attack surface

Finally, the potential severity of different attack scenarios will have to be evaluated and compared. For this process, an established method such as the Common Vulnerability Scoring System (CVSS) can be used. CVSS uses a score from zero to ten to help rank different attack scenarios. An example is given in fig. 25.7.

		HMI	Firmware	WLAN	Backend Applications	EAI	Enterprise Applications
S	Spoofing identity	●	●	●	●	●	●
T	Tampering with data		●		●		●
R	Repudiation	●		●		●	
I	Information disclosure		●		●	●	
D	Denial of service		●	●			●
E	Elevation of privilege			●	●		

● CVSS critical: 9.0-10   ● CVSS high: 7.0-8.9   ● CVSS medium: 4.0-6.9

Fig. 25.7 CVSS

Next, the product team needs to define a set of criteria for dealing with the risks on the different levels, e.g.

- High risk: Fixed immediately
- Medium risk: Fixed in next minor release
- Low risk: Fixed in next major release

To manage the identified and classified risks, a risk catalog or risk register is created to track the risks and the status. This would usually be done as part of the overall defect tracking.

25.6.1   Security Architecture & Setup

Securing an AIoT system is not a single task, and the results of the threat modeling exercise are likely to show attack scenarios of very different kinds. Some of these scenarios will have to be addressed during the later phases of the DevSecOps cycle, e.g., during development and testing. However, some basic security measures can usually already be established as part of the system architecture and setup, including:

- Basic security measures, such as firewalls and anti-virus software
- Installation of network traffic monitors and port scanners
- Hardware-related security architecture measures, e.g., Trusted Platform Module (TPM) for extremely sensitive systems

These types of security-related architecture decisions should be made in close alignment with the product architecture team, early in the architecture design.

25.6.2   Integration, Testing, and Operations

In **DevSecOps**, the development teams must be integrated into all security-related activities. On the code-level, regular code reviews from a security perspective can be useful. On the hardware-level, design and architecture reviews should be performed from a security perspective as well. For AI, the actual coding is usually only a small part of the development. Model design and training play a more important role and should also be included in regular security reviews.

Continuous Integration has to address security concerns specifically on the code level. Code-level security tests/inspections include:

- Before compilation/packaging: SAST can be used for Static Application Security Testing.
- IAST (Interactive application security testing) uses code instrumentation, which can slow down performance. Individual decisions about enabling/disabling it will have to be made as part of the CI process.

Security testing includes tests with a specific focus on testing for security vulnerabilities. These can include:

- Applications, e.g., DAST (Dynamic Application Security Testing)
- Hardware-related security tests
- AI model security tests
- End-to-End System, e.g., manual and automated penetration tests

Secure operations have to include a number of activities, including:

- Threat Intelligence
- Infrastructure and Network Testing (including Secure OTA)
- Security tests in the field
- RASP: Runtime Application Self-Protection
- Monitor/Detect/Response/Recover

25.6.3   Minimum Viable Security

The key challenge with security planning and implementation is to find the right approach and the right level of required resource investments. If too little attention (and % of project resources and budget) is given to security, then there is a good chance that this will result in a disaster - fast. However, if the entire project is dominated by security, this can also be a problem. This relates to the resources allocated to different topics, but also to the danger of over-engineering the security solutions (and in the process making it too difficult to deliver the required features and usability). Figuring out the Minimum Viable Security is something that must be done between product management and security experts. Also, it is important that this is seen as an ongoing effort, constantly reacting to new threats and supporting the system architecture as it evolves.

25.7   Trust Policy Management for AIoT

In addition to security-related activities, an AIoT product team should also consider taking a proactive approach toward broader trust policies. These trust policies can include topics such as:

- Data sharing policies (e.g., sharing of IoT data with other stakeholders)
- Transparency policies (e.g., making data sharing policies transparent to end users)
- Ethics-related policies (e.g., for AI-based decisions)

Taking a holistic view of AIoT trust policies and establishing central trust policy management can significantly contribute to creating trust between all stakeholders involved.

Chapter 26
Reliability & Resilience

Dirk Slama

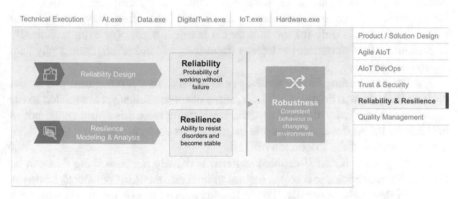

Fig. 26.1 Ignite AIoT - Reliability & Resilience

Ensuring a high level of robustness for AIoT-based systems is usually a key requirement. Robustness is a result of two key concepts: reliability and resilience ("R&R"). Reliability concerns designing, running and maintaining systems to provide consistent and stable services. Resilience refers to a system's ability to resist adverse events and conditions (Fig. 26.1).

Ensuring reliability and resilience is a broad topic that ranges from basics such as proper error handling on the code level up to georeplication and disaster recovery. In addition, there are some overlaps with Security, as well as Verification and Validation. This chapter first discusses reliability and resilience in the context of AIoT DevOps and then looks at the AI and IoT specifics in more detail.

D. Slama (✉)
Ferdinand Steinbeis Institute, Berlin, Germany
e-mail: dirk.slama@bosch.com

26.1 R&R for AIoT DevOps

Traditional IT systems have been using reliability and resilience engineering methods for decades. The emergence of hyperscaling cloud infrastructures has taken this to new levels. Some of the best practices in this space are well documented, for example, Google's Site Reliability Engineering approach for production systems [24]. These types of systems need to address challenges such as implementing recovery mechanisms for individual IT services or entire regions, dealing with data backups, replication, clustering, network load-balancing and failover, georedundancy, etc.

The IoT adds to these challenges because parts of the system are implemented not in the data center but rather as hardware and software components that are deployed in the field. These field deployments can be based on sophisticated EDGE platforms or on some very rudimentary embedded controllers. Nevertheless, IT components deployed in the field often play by different rules — and if it is only for the fact that it is much harder (or even technically or economically impossible) to access them for any kind of unplanned physical repairs or upgrades.

Finally, AI is adding further challenges in terms of model robustness and model performance. As will be discussed later, some of these challenges are related to the algorithmic complexity of the AI models, while many more arise from complexities of handling the AI development cycle in production environments, and finally adding the specifics of the IoT on top of it all.

Ensuring R&R for AIoT-enabled systems is usually not something that can be established in one step, so it seems natural to integrate the R&R perspective into the AIoT DevOps cycle. Naturally, the R&R perspective must be integrated with each of the four AIoT DevOps quadrants. From the R&R perspective, agile development must address not only the application code level but also the AI/model level, as well as the infrastructure level. Continuous Integration must ensure that all R&R-specific aspects are integrated properly. This can go as far as preparing the system for Chaos Engineering Experiments (Fig. 26.2) [25]. Continuous Testing must ensure that all R&R concepts are continuously validated. This must include basic system-level R&R, as well as AI and IoT-specific R&R aspects. Finally, Continuous Delivery/Operations must bring R&R to production. Some companies are even going to the extreme to conduct continuous R&R tests as part of their production systems (one big proponent of this approach is Netflix, where the whole Chaos Engineering approach originated).

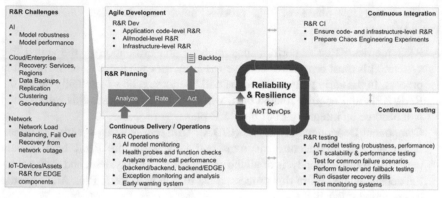

Fig. 26.2   R&R DevOps for AIoT

26.1.1   R&R Planning: Analyze Rate Act

While it is important that R&R is treated as a normal part of the AIoT DevOps cycle, it usually makes sense to have a dedicated R&R planning mechanism, which looks at R&R specifically. Note that a similar approach has also been suggested for Security, as well as Verification & Validation. It is important that none of these three areas is viewed in isolation, and that redundancies are avoided.

The AIoT Framework proposes a dedicated Analyze/Rate/Act planning process for R&R, embedded into the AIoT DevOps cycle, as shown in Fig. 26.3.

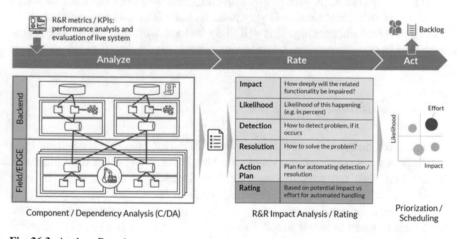

Fig. 26.3   Analyze Rate Act

The "Analyze" phase of this process must take two key elements into consideration:

- R&R metrics/KPIs: A performance analysis and evaluation of the actual live system. This must be updated and used as input for each iteration of the planning process. In the early phases, the focus will be more on how to actually define the R&R KPIs and acquire related data, while in the later phases this information will become an integral part of the R&R planning process.
- Component/Dependency Analysis (C/DA): Utilizing existing system documentation such as architecture diagrams and flowcharts, the R&R team should perform a thorough analysis of all the components in the system, and their potential dependencies. From this process, a list of potential R&R Risk Areas should be compiled ("RA list").

The RA list can contain risks at different levels of granularity, ranging from risks related to the availability of individual microservices up to risks related to the availability of entire regions. The RA list must also be compared to the results of the Threat Modeling that comes out of the DevSecOps planning process. In some cases, it can even make sense to join these two perspectives into a single list or risk repository.

The "Rate" phase must look at each item from the RA list in detail, including the potential impact of the risk, the likelihood that it occurs, ways of detecting issues related to the risk, and ways for resolving them. Finally, a brief action plan should describe a plan for automating the detection and resolution of issues related to the risk, including a rough effort estimate. Based on all of the above, a rating for each item in the RA list should be provided.

The "Act" phase starts with prioritizing and scheduling the most pressing issues based on the individual ratings. Highly rated issued must then be transferred to the general development backlog. This will likely include additional analysis of dependencies to backlog items more related to the application development side of things.

26.1.2 Minimum Viable R&R

Similar to the discussion on Minimum Viable Security, project management must carefully strike a balance between investments in R&R and other investments. A system that does not support basic R&R will quickly frustrate users or even worse — result in lost business or actual physical harm. However, especially in the early stages of market exploration, the focus must be on features and usability. Determining when to invest in R&R as the system matures is a key challenge for the team.

26.2 Robust, AI-Based Components in AIoT

The AI community is still in the early stages of addressing reliability, resilience and related topics such as robustness and explainability of AI-based systems. H. Truong provides the following definitions (Fig. 26.3) [26] from the ML perspective:

- Robustness: Dealing with imbalanced data and learning in open-world(out-of-distribution) situations
- Reliability: Reliable learning and reliable inference in terms of accuracy and reproducibility of ML models; uncertainties/confidence in inferences; reliable ML service serving
- Resilience: bias in data, adversary attacks in ML, resilience learning, computational Byzantine failures

In the widely cited paper on *Hidden Technical Debt in Machine Learning Systems* (Fig. 26.4) [27], the authors emphasize that only a small fraction of real-world ML systems are composed of ML code, while the required surrounding infrastructure is vast and complex, including configuration, data collection, feature extraction, data verification, machine resource management, analysis tools, process management tools, serving infrastructure, and monitoring.

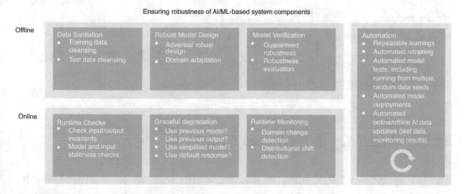

Fig. 26.4 Robust AI Components for AIoT

The AIoT Framework suggests differentiating between the online and offline perspectives of the AI-based components in the AIoT system. The offline perspective must cover data sanitation, robust model design, and model verification. The online perspective must include runtime checks (e.g., feature values out of range or invalid outputs), an approach for graceful model degradation, and runtime monitoring. Between the online and offline perspectives, a high level of automation must be achieved, covering everything from training to testing and deployments.

Mapping all of the above R&R elements to an actual AIoT system architecture is not an easy feat. Acquiring high-quality test data from assets in the field is not always easy. Managing the offline AI development and experimentation cycle can rely on standard AI engineering and automation tools. However, model deployments to assets in the field rely on nonstandard mechanisms, e.g., relying on OTA (over-the-air) updates from the IoT toolbox. Dealing with multiple instances of models deployed onto multiple assets (or EDGE instances) in the field is something that goes beyond standard AI processing in the cloud. Finally, gathering — and making sense of — monitoring data from multiple instances/assets is beyond today's well-established AI engineering principles (Fig. 26.5).

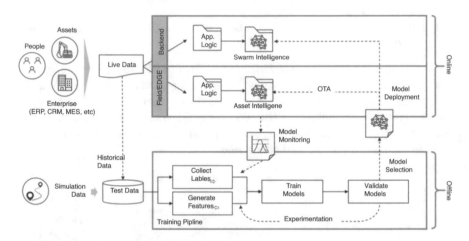

Fig. 26.5 Architecture for robust, AI-enabled AIoT components

26.2.1   Reliability & Resilience for IoT

The IoT specifics of Reliability & Resilience also need to be addressed. For the backend (cloud or enterprise), of course, most of the standard R&R aspects of Internet/cloud/enterprise systems apply. Since the IoT adds new categories of clients (i.e., assets) to access the back ends, this has to be taken into consideration from an R&R perspective. For example, the IoT backend must be able to cope with malfunctioning or potentially malicious behaviour of EDGE or embedded components.

For the IoT components deployed in the field, environmental factors can play a significant role, which requires extra ruggedness for hardware components, which can be key from the R&R perspective. Additionally, depending on the complexity of the EDGE/embedded functions, many of the typical R&R features found in modern cloud environments will have to be reinvented to ensure R&R for components deployed in the field.

Finally, for many IoT systems — especially where assets can physically move — there will be much higher chances of losing connectivity from the asset to the backend. This typically requires that both backend and field-based components implement a certain degree of autonomy. For example, an autonomous vehicle must be able to function in the field without access to additional data (e.g., map data) from the cloud. Equally, a backend asset monitoring solution must be able to function, even if the asset is currently not connected. For example, asset status information must be augmented with a timestamp that indicates when this information was last updated (Fig. 26.6).

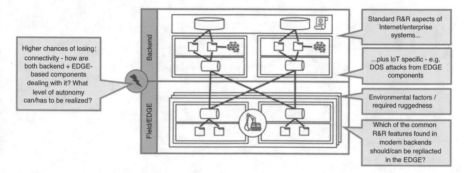

Fig. 26.6 Building robust IoT solutions

Chapter 27
Quality Management

Dirk Slama

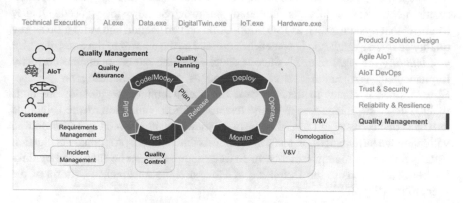

Fig. 27.1 Ignite AIoT - Artificial Intelligence

Quality Management (QM) is responsible for overseeing all activities and tasks needed to maintain a desired level of quality. QM in Software Development traditionally has three main components: quality planning, quality assurance, and quality control. In many agile organizations, QM is becoming closely integrated with the DevOps organization. Quality Assurance (QA) is responsible for setting up the organization and its processes to ensure the desired level of quality. In an agile organization, this means that QA needs to be closely aligned with DevOps. Quality Control (QC) is responsible for the output, usually by implementing a test strategy along the various stages of the DevOps cycle. Quality Planning is responsible for setting up the quality and test plans. In a DevOps organization, this will be a continuous process (Fig. 27.1).

D. Slama (✉)
Ferdinand Steinbeis Institute, Berlin, Germany
e-mail: dirk.slama@bosch.com

© The Author(s) 2023
D. Slama et al. (eds.), *The Digital Playbook*,
https://doi.org/10.1007/978-3-030-88221-1_27

QM for AIoT-enabled systems must take into consideration all the specific challenges of AIoT development, including QM for combined hardware/software development, QM for highly distributed systems (including edge components in the field), as well as any homologation requirements of the specific industry. Verification & Validation (V&V) usually plays an important role as well. For safety relevant systems (e.g., in transportation, aviation, energy grids), Independent Verification & Validation (IV&V) via an independent third party can be required.

27.1 Verification & Validation

Verification and validation (V&V) are designed to ensure that a system meets the requirements and fulfills its intended purpose. Some widely used Quality Management Systems, such as ISO 9000, build on verification and validation as key quality enablers. Validation is sometimes defined as the answer to the question *"Are you building the right thing?"* since it checks that the requirements are correctly implemented. Verification can be expressed as *"Are you building the product right?"* since it relates to the needs of the user. Common verification methods include unit tests, integration tests and test automation. Validation methods include user acceptance tests and usability tests. Somewhere in between verification and validation we have regression tests, system tests and beta test programs. Verification usually links back to requirements. In an agile setup, this can be supported by linking verification tests to the Definition of Done and the Acceptance Criteria of the user stories (Fig. 27.2).

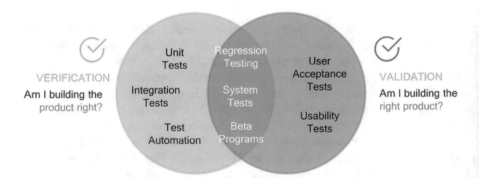

Fig. 27.2 Quality Control

27.2   Quality Assurance and AIoT DevOps

So how does Quality Assurance fit with our holistic AIoT DevOps approach? First, we need to understand the quality-related challenges, including functional and non-functional. Functional challenges can be derived from the agile story map and sprint backlogs. Non-functional challenges in an AIoT system will be related to AI, cloud and enterprise systems, networks, and IoT/edge devices. In addition, previously executed tests, as well as input from ongoing system operations, must be taken into consideration. All of this must serve as input to the Quality Planning. During this planning phase, concrete actions for QA-related activities in development, integration, testing and operations will be defined.

QA tasks during development must be supported both by the development team, and by any dedicated QA engineers. The developers usually perform tasks such as manual testing, code reviews, and the development of automated unit tests. The QA engineers will work on the test suite engineering and automation setup.

During the CI phase (Continuous Integration), basic integration tests, automated unit tests (before the check-in of the new code), and automatic code quality checks can be performed.

During the CT phase (Continuous Testing), many automated tests can be performed, including API testing, integration testing, system testing, automated UI tests, and automated functional tests.

Finally, during Continuous Delivery (CD) and operations, User Acceptance Test (UATs) and lab tests can be performed. For an AIoT system, digital features of the physical assets can be tested with test fleets in the field. Please note that some advanced users are now even building test suites that are embedded with the production systems. For example, Netflix became famous for the development of the concept of chaos engineering. By letting loose an "army" of so-called Chaos Monkeys onto their production systems, they forced the engineers to ensure that their systems withstand turbulent and unexpected conditions in the real world. This is now referred to as "Chaos Engineering" (Fig. 27.3).

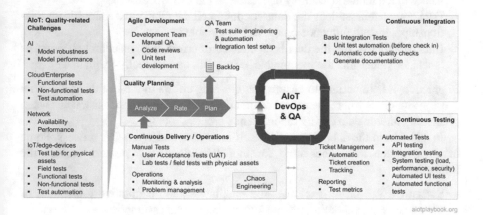

Fig. 27.3 Quality Assurance and AIoT DevOps

27.3   Quality Assurance for AIoT

What are some of the AIoT-specific challenges for QA? The following looks at QA
& AI, as well as the integration perspective. AI poses its own set of challenges on
AI. And the integration perspective is important since an AIoT system, by its very
nature, will be highly distributed and consist of multiple components.

27.3.1   QA & AI

QA for AI has some aspects that are very different from traditional QA for software.
The use of training data, labels for supervised learning, and ML algorithms instead
of code with its usual IF/THEN/ELSE-logic poses many challenges from the QA
perspective. The fact that most ML algorithms are not "explainable" adds to this.

From the perspective of the final system, QA of the AI-related services usually
focuses on functional testing, considering AI-based services a black box ("Black
Box Testing") which is tested in the context of the other services that make up the
complete AIoT system. However, it will usually be very difficult to ensure a high
level of quality if this is the only test approach. Consequently, QA for AI services in
an AIoT system also requires a "white box" approach that specifically focuses on
AI-based functionality.

In his article "Data Readiness: Using the 'Right' Data" (Sect. 27.3.1) [28], Alex
Castrounis describes the following considerations for the data used for AI models:

- Data quantity: does the dataset have sufficient quantity of data?
- Data depth: is there enough varied data to fill out the feature space (i.e., the num-
  ber of possible value combinations across all features in a dataset)?
- Data balance: does the dataset contain target values in equal proportions?
- Data representativeness: Does the data reflect the range and variety of feature
  values that a model will likely encounter in the real world?
- Data completeness: does the dataset contain all data that have a significant rela-
  tionship with and influence on the target variable?
- Data cleanliness: has the data been cleaned of errors, e.g., inaccurate headers or
  labels, or values that are incomplete, corrupted, or incorrectly formatted?

In practice, it is important to ensure that cleaning efforts in the test dataset are not
causing situations where the model cannot deal with errors or inconsistencies when
processing unseen data during the inference process.

In addition to the data, the model itself must also undergo a QA process. Some
of the common techniques used for model validation and testing include the
following:

- **Statistical validation** examines the qualitative and quantitative foundation of
  the model, e.g., validating the model's mathematical assumptions

- The **holdout method** is a basic type of cross-validation. The dataset is split into two sets, the training set and the test set. The model is trained on the training set. The test set is used as "unseen data" to evaluate the skill of the model. A common split is 80% training data and 20% test data.
- **Cross-validation** is a more advanced method used to estimate the skill of an ML model. The dataset is randomly split into k "folds" (hence "k fold cross-validation"). One fold is used as the test set, the k-1 for training. The process is repeated until each fold has been used once as the test set. The results are then summarized with the mean of the model skill scores.
- **Model simulation** embeds the final model into a simulation environment for testing in near-real-world conditions (as opposed to training the model using the simulation).
- **Field tests** and **production tests** allow for testing of the model under real-world conditions. However, for models used in functional safety-related environments, this means that in the case of badly performing models, a safe and controlled degradation of the service must be ensured (Fig. 27.4).

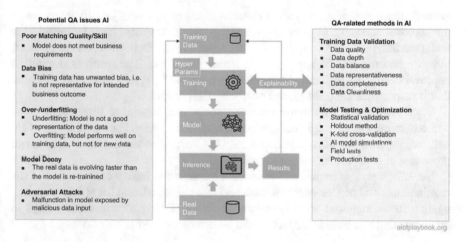

Fig. 27.4   QA for AI

27.3.2   Integrated QA for AIoT

At the service level, AI services can usually be tested using the methods outlined in the previous section. After the initial tests are performed by the AI service team, it is important that AI services be integrated into the overall AIoT product for real-world integration tests. This means that AI services are integrated with the remaining IoT services to build the full AIoT system. This is shown in Fig. 27.5. The fully integrated system can then be used for User Acceptance Tests, load and scalability tests, and so on.

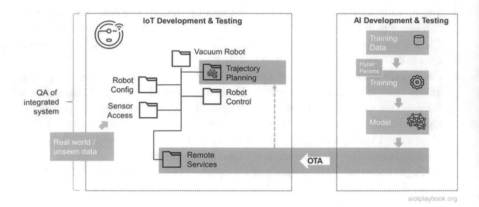

Fig. 27.5 QA for AIoT

27.4 Homologation

Usually, the homologation process requires the submission of an official report to the approval authority. In some cases, a third-party assessment must be included as well (see Independent Verification & Validation above). Depending on the product, industry and region, the approval authorities will differ. The result is usually an approval certificate that can either relate to a product ("type") or the organization that is responsible for creating and operating the product.

Since AIoT combines many new and sometimes emerging technologies, the homologation process might not always be completely clear. For example, there are still many questions regarding the use of OTA and AI in the automotive approval processes of most countries.

Nevertheless, it is important for product managers to have a clear picture of the requirements and processes in this area, and that the foundation for efficient homologation in the required areas is ensured early on. Doing so will avoid delays in approvals that can have an impact on the launch of the new product (Fig. 27.6).

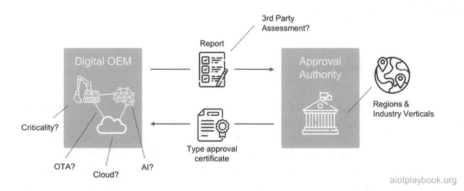

Fig. 27.6 AIoT and Homologation

Part V
Case Studies

Part V of the Digital Playbook provides a number of case studies to illustrate many of the concepts discussed thus far in more detail and from a real-world application perspective. The first case study, from Bosch and Microsoft, looks at AIoT in a global, high-volume manufacturing network. This is a good example of a "harvesting" type of AIoT organization, as introduced in the AIoT long-tail discussion. The second case study, from TÜV SÜD, looks at Drone-based facade inspection. The third case study, from Bosch Rexroth, is looking at Predictive Maintenance for Hydraulic Components. The last case study examines how Bürkert addresses the long tail of AIoT using the BaseABC method.

Each of these case studies highlights different aspects of AIoT. Some of them are more AIoT Short Tail opportunities, while others are more Long Tail opportunities. Figure 1 shows their positions relative to each other. The large hadron collider case study is not included in this book rendition of the *Digital Playbook*. It can be found online here. The vacuum robot example is discussed in detail in the AIoT Product / Solution Design section.

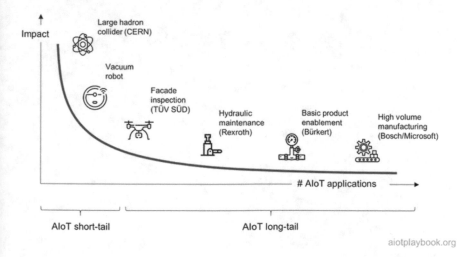

Fig. 1 Long and Short Tail of AIoT Case Studies

Chapter 28
AIoT in a Global, High-Volume Manufacturing Network (Bosch and Microsoft)

Dirk Slama

28.1 Introduction

Bosch Chassis Systems Control (CC) is a division of Bosch that develops and manufactures components, systems and functions in the field of vehicle safety, vehicle dynamics and driver assistance. The products from Bosch CC combine cameras, radar and ultrasonic sensors, electric power steering and active or passive safety systems to improve driver safety and comfort. Bosch CC is a global organization with twenty factories around the world. Very high volumes, combined with a high product variety, characterize Bosch CC production. Large numbers of specially designed and commissioned machines are deployed to ensure high levels of automation. Organized in a global production network, plants can realize synergies at scale (Fig. 28.1).

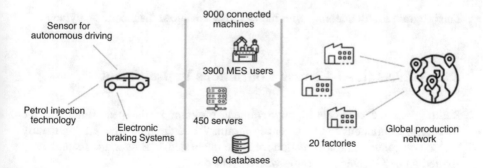

Fig. 28.1 Overview Bosch CC Case Study

D. Slama (✉)
Ferdinand Steinbeis Institute, Berlin, Germany
e-mail: dirk.slama@bosch.com

© The Author(s) 2023
D. Slama et al. (eds.), *The Digital Playbook*,
https://doi.org/10.1007/978-3-030-88221-1_28

Naturally, IT plays an important role in product engineering, process development, and manufacturing. Bosch CC has more than 90 databases, 450 servers, and 9000 machines connected to them. A total of 3900 users are accessing the central MES system (Manufacturing Execution System) to track and document the flow of materials and products through the production process.

28.2 Phase I: Data-Centric Continuous Improvement

Like most modern manufacturing organizations, Bosch CC strives to continually improve product output and reduce costs while maintaining the highest levels of quality. One of the key challenges of doing this in a global product network is to standardize and harmonize. The starting point for this is actually the machinery and equipment in the factories, especially the data that can be accessed from it. Over the years, Bosch CC has significantly invested in such harmonization efforts. This was the prerequisite for the first data-centric optimization program, which focused on EAI (Enterprise Application Integration, with a focus on data integration), as well as BI (Business Intelligence, with a focus on data visualization). Being able to make data in an easily accessible and harmonized way available to the production staff resulted in a 13% output increase per year in the last five years. The data-centric continuous improvement program was only possible because of the efforts in standardizing processes, machines and equipment and making the data for the 9000 connected machines easily accessible to staff on the factory floor. The data-centric improvement initiative mainly focuses on two areas:

- Descriptive Analytics (Visual Analytics): *What happened?*
- Diagnostic Analytics (Data Mining): *Why did it happen?*

The program is still ongoing, with increasingly advanced diagnostic analytics.

28.3 Phase II: AI-Centric Continuous Improvement

Building on the data-centric improvement program is the next initiative, the AI-centric Continuous Improvement Program. While the first wave predominantly focused on gaining raw information from the integrated data layer, the second wave applied AI and Machine Learning with a focus on:

- Predictive Analytics (ML): *What will happen?*
- Prescriptive Analytics (ML): *What to do about it?*

As of the time of writing, this new initiative has been going on for two years, with the first results coming out of the initial use cases, supporting the assumption that this initiative will add at least a further 10% in annual production output increase (Fig. 28.2).

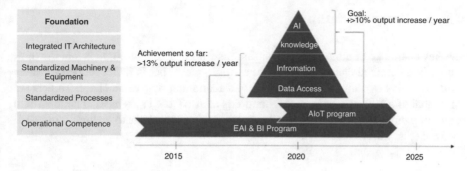

Fig. 28.2 Bosch CC Case Study – Phases

28.4   Closed-Loop Optimization

The approach taken by the Bosch CC team fully supports the Bosch AIoT Cycle, which assumes that AI and IoT support the entire product lifecycle, from product design over production setup to manufacturing. The ability to gain insights into how products are performing in the field provides an invaluable advantage for product design and engineering. Closing the loop with machine building and development departments via AI-gained insights enables the creation of new, more efficient machines, as well as product designs better suited to efficient production. Finally, applying AI to machine data gained via the IoT enables root cause analysis for production inefficiencies, optimization of process conditions, and bad part detection and machine maintenance requirement predictions (Fig. 28.3).

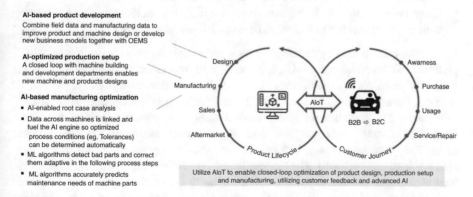

Fig. 28.3 Bosch CC: Closed-Loop Optimization

28.5   Program Setup

A key challenge of the AI-driven Continuous Improvement program is that the optimization potential cannot be found in a single place, but is rather hidden in many different places along the engineering and manufacturing value chain. To achieve the goal of a 10% output increase, hundreds of AIoT use cases have to be identified and implemented every year. This means that this process has to be highly industrialized (Fig. 28.4).

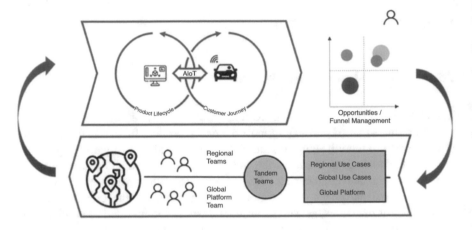

Fig. 28.4 Bosch CC AIoT Program Setup

To set up such an industrialized approach, the program leadership identified a number of success factors, including:

- Establishment of a global team to coordinate the efforts across all factories in the different regions and provide centralized infrastructure and services
- Close collaboration with the experts in the different regions, working together with experts from the global team in so-called tandem teams

The global team started by defining a vision and execution plan for the central AIoT platform, which combines an AI pipeline with central cloud compute resources as well as edge compute capabilities close to the lines on the factory floors.

Next, the team started to work with the regional experts to identify the most relevant use cases. Together, the global team and regional experts prioritize these use cases. The central platform is then gradually advanced to support the selected use cases. This ensures that the platform features always support the needs of the use case implementation teams. The tandem teams consist of central platform experts as well as regional process experts. Depending on the type of use case, they include Data Analysts, and potentially Data Scientists for the development of more complex models. Data Engineers support the integration of the required systems, as well as potentially required customization of the AI pipeline. The teams strive to ensure that the regionally developed use cases are integrated back into the global use case portfolio so that they can be made accessible to all other factories in the Bosch CC global manufacturing network.

28.6 AIoT Platform and AI Pipeline

The AIoT platform being built by Bosch CC combines traditional data analytics capabilities with advanced AI/ML capabilities. The data ingest layer integrates data from all relevant data sources, including MES and ERP. Both batch and real-time ingest are supported. Different storage services are available to support the different input types. The data analytics layer is running on Microsoft Azure, utilizing Tableau and Power BI for visual analytics. For advanced analytics, a machine learning framework is provided, which can utilize dedicated ML compute infrastructure (GPUs and CPUs), depending on the task at hand. The trained models are stored in a central model repository. From there, they can be deployed to the different edge nodes in the factories. Local model monitoring helps to gain insights into model performance and support alerting. The AI pipeline supports an efficient CI/CD process and allows for automated model retraining and redeployment (Fig. 28.5).

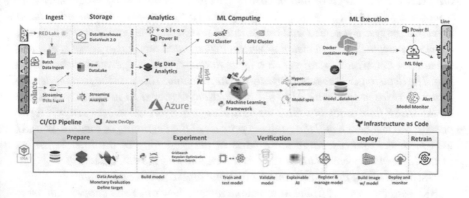

Fig. 28.5 Bosch CC AIoT Architecture

28.7 Expert Opinions

In the following interview with Sebastian Klüpfel (Bosch CC central AI platform team) and Uli Homann (Corporate Vice President at Microsoft and a member of the AIoT Editorial Board), some insights and lessons learned were shared.

Dirk Slama: *Sebastian, how did you get started with your AI initiative?*

Sebastian Klüpfel: *Since 2000, we have been working on the standardization and connection of our manufacturing lines worldwide. Our manufacturing stations provide a continuous flow of data. For each station, we can access comprehensive information about machine conditions, quality-relevant data and even individual sensor values. By linking the upstream and downstream production stages on a data side, we achieve a perfect vertical connection. For example, for traceability reasons we store 2500 data points per ABS/ESP part (ABS: Anti-Locking Brake, ESP: Electronic Stability Program). We use these data as the basis for our continuous improvement process. Thus, in the ABS/ESP manufacturing network, we were able to increase the production rate by 13% annually over the last 18 years. All of this was accomplished just by working on the lower part of the I4.0 pyramid: the data access/information layer.*

DS: *Could you please describe the ABS/ESB use case in more detail?*

SK: *This was our first use case. Our first use case for AI in manufacturing is at the end of the ABS/ESP final assembly, where all finished parts are checked for functionality. To date, during this test process, each part has been handled and tested separately. Thus, the test program did not recognize whether a cause for a bad test result was a real part defect or just a variation in the testing process. As a result, we started several repeated tests on bad tested parts to ensure that the part was faulty. A defective was thus tested up to four times until the final result was determined. This reduced the output of the line because the test bench was operating on its full capacity by repeated tests. By applying AI, we reduce the bottleneck at the test bench significantly. Based on the first bad test result, the AI can detect whether a repeated test is useful or not. In case of variations in the testing process, we start a second test, and thus, we save the right part as a "good part". On the other hand, defective parts are detected and immediately rejected. Thus, unnecessary repeated checks are avoided. The decision of whether a repeated test makes sense is made by a neuronal net, which is trained on a large database and is operated closely to the line. The decision is processed directly on the machine. This was the first implementation of a closed loop with AI in our production. Our next AI use case was also in the ABS/ESP final assembly. Here we have recognized a relation between bad tested parts at the test bench and the caulking process at the beginning of the line. With AI, we can detect and discharge these parts at the beginning of the line, before we install high-quality components such as the engine and the control unit.*

DS: *What about the change impact on your organization and systems?*

SK: *With the support from our top management, new roles and cooperation models were established. The following roles were introduced and staffed to implement AI use cases:*

- *Leadership: management and domain leaders need to understand strategic relevance, advantages, limits, and an overview of tools.*
- *Citizen Data Scientist: work on an increasingly data-driven field and uses analytical tools applied to its domain. Therefore, a basic understanding and knowledge of Big Data and machine learning is necessary.*
- *Data Engineer: builds Big Data systems and knows how to connect to these systems and machine learning tools. Therefore, deep knowledge of IT systems and development is necessary.*

- *Data Scientist: develops new algorithms and methods with deep knowledge in existing methods. Therefore, the data scientist must be up-to-date and have know-how in the CRISP-DM analytics project lead.*

We want to use the acquired data in combination with AI with the maximum benefit for the profitability of our plant (data-driven organization). Only through the interaction and change in the working methods of people, machines and processes/organization can we create fundamentally new possibilities to initiate improvement processes and achieve productivity increases. We rely on cross-functional teams from different domains to guarantee quick success.

DS: *What were the major challenges you were facing?*

SK: *For the implementation of these and future AI use cases, we rely on a uniform architecture. Without this architecture, the standardized and industrialized implementation of AI use cases is not possible. The basis is the detailed data for every manufacturing process, which are acquired from our standardized and connected manufacturing stations. Since 2000, we have implemented an MES (Manufacturing Execution System) that enables holistic data acquisition. The data is stored in our cloud (Data Lake). As a link between our machines and the cloud, we use proven web standards. After an intensive review of existing cloud solutions internally and externally, we decided to use the external Azure Cloud from Microsoft. Here, we can use as many resources as we need for data storage, training of AI models and preprocessing of data (Data Mart). We also scale financially, and we only create costs where we have a benefit. Thus, we can also offer the possibility to analyze the prepared data of our Data Mart via individually created evaluations and diagrams (Tableau, PowerBI). We run our trained models in an edge application close to our production lines. By using this edge application, we bring the decisions of the AI back to the line. For the connection of the AI to the line, only minimal adjustments to the line are necessary, and we guarantee a fast transfer of new use cases to other areas.*

DS: *How does your ROI look like for the first use cases?*

SK: *Since an AI decides on repeated tests on the testing stations of the ABS/ESP final assembly, we can detect 40% of the bad parts after the first test cycle. Before introducing the AI solution, the bad parts always went through four test cycles. Since the test cells are the bottleneck stations on the line, the saved test cycles can increase the output of the line, reduce cycle time, increase quality and reduce error costs. This was proven on a pilot line. The rollout of the AI solution offers the potential for an increase in output of approximately 70,000€ per year and a cost reduction of nearly one million (since no additional test stations are needed to be purchased for more complex testing). This was only the beginning. With our standardized architecture, we have the foundation for a quick and easy implementation of further AI use cases. By implementing AI in manufacturing, we expect an increase in productivity of 10% in the next five years.*

DS: *What are the next steps for you?*

SK: *Our vision is that in the future, we will fully understand all cause-effect relations between our product, machine and processes and create a new way of learning with the help of artificial intelligence to assist our people in increasing the productivity of our lines. To achieve this, the following steps are planned and are already in progress:*

- *Pioneering edge computing: First, we are working on a faster edge application. We have to bring the decisions from AI even faster to the manufacturing station. For the first use cases, our edge-application is still sufficient. However, for AI use cases for short-cycle assembly lines (approximately 1 second cycle time), the actual edge solution is no longer sufficient. Here, we are already working on solutions to deliver the predictions back to the line within fractions of a second even for such use cases.*
- *Automated machine learning: 80% of our data are already preprocessed automatically. Our target is to further increase the automation rate. In addition, we have ideas how to automate the selection of the right ML model with an appropriate hyper parameter search. Of course, we are also working on an automated analysis of ML decisions to monitor the health status of models in production.*
- *Implement more use cases: Our architecture is designed for thousands of AI use cases. We have to identify and implement these. By doing so, we ensure that we do not implement show cases. We want to implement real use cases for our Digital Factory, including:*

  - *Predict process parameters: learn optimal process parameters from prior processes (e.g. prior to final assembly)*
  - *Adaptive tolerances*
  - *Bayesian network: We want to train a Bayesian network on all parameters of the HU9 final assembly. This means that influences and relations can be read from the graph. Relations are much deeper than pairwise correlations.*

DS: *What are the key lessons learned thus far?*

SK: *For a continuous improvement program like ours, KI must be industrialized: we have been proven right with our approach to make AI applicable on a large scale instead of individual "lighthouse projects", which are not easily adaptable to other use cases. Another key success factor is the standardized architecture: by storing all the data in a cloud, we can ensure a holistic view of the data across all factories in our global manufacturing network and use this view to train the models centrally with all the required data at hand. Next, we can take the trained models and deploy them on the edge layer as close to the actual production lines as possible. To work quickly and efficiently on the implementation of AI use cases, new competence and work models must be established. The most important thing is that digital transformation must be a part of corporate strategy. After all, digital transformation can succeed only if all employees work together toward the path of a data-driven organization.*

DS: *Uli, what is your perspective on this?*

Uli Homann: *In this project, Bosch shows first-hand how to streamline operations, cut costs, and accelerate innovation across the entire product life-cycle by driving change holistically across culture, governance, and technology. Through partnership with Microsoft, Bosch is unlocking the convergence of OT and IT. The holistic approach provided by the* Digital Playbook *enables a continuous feedback loop that helps teams turn AI-driven data insights into business value at scale.*

DS: *Thank you, both!*

Chapter 29
Drone-Based Facade Inspection (TUEV SUED)

Dirk Slama

This case study describes a drone-based system for automated building façade inspection that utilizes AIoT for drone management and image-analytics. The system was developed by the Real Estate & Infrastructure Division of TÜV SÜD.

29.1 Building Façades and Related Challenges

Building façades are an important aspect of buildings, both from an architectural as well as from an engineering perspective. Building façades have a huge impact not only on aesthetics but also on energy efficiency and safety. Especially in high-rise buildings, the façade can be quite complex, combining a number of different materials, including concrete, glass, steel, polymers and complex material mixes.

Problems with building façades can arise during construction as well as during the building operations phase. Typical problems include cracks in different materials, concrete spalling, corrosion, delamination, decolorization, efflorescence, peeling and flaking, chalking, hollowness, sealant deterioration, and so on. While some of these problems only have an impact on the optics of the buildings, others can have a quite severe impact on safety, e.g., because of façade elements falling down from high heights, increased risk of fires, or even complete collapses.

D. Slama (✉)
Ferdinand Steinbeis Institute, Berlin, Germany
e-mail: dirk.slama@bosch.com

29.2   Façade Inspection

Façade inspection is an integral part of building maintenance, especially for high-rise buildings. It helps to verify the integrity of the building structure and ensures safety for its occupants and people passing by. However, conventional manual façade inspection can be time, labor and cost intensive, and disruptive for building occupants, and dangerous for inspectors due to difficult access at height. Finally, the results of manual façade inspection can be subjective, depending on the expertise of the inspector.

In some countries, regular façade inspections are required by regulators. Regulations usually differ depending on building size and age. For example, in Singapore, buildings older than 20 years old and over 13 meters in height have to undergo façade inspections every 7 years. In other countries, the requirements for periodic façade inspections are driven more by building insurance companies.

29.3   Automated Façade Inspection

Automated façade inspection solutions must accurately scan the exterior of buildings, e.g., utilizing drones to carry high-resolution cameras. The Smart Façade Inspection service of TÜV SÜD caters to building owners and operators of large high-rise buildings and helps construction companies ensure façade quality and monitor construction progress.

29.3.1   Customer Journey

The customer journey of the automated façade inspection solution starts with the customer request for the service. Based on the customer information provided, the service operator (TÜV SÜD) will prepare the required documentation and apply with the required authorities for drone flight approvals. On-site inspection will be carried out by a specialized drone operations team. The data, inspection results and a 3D model of the façade will be made available via a specialized cloud platform.

29.3.2   Customer Benefits

Customer benefits include:

- The results are available in a fraction of the time compared to conventional inspection

- Digital representation of the façade and whole building facilitates building operation
- Automated digital workflow and data benchmarking improve service quality and interoperability
- Domain experts for standards and best practice, ensuring up-to-date compliance to continually evolving regulations

29.4 Implementation with AIoT

At the core of the operational system is a smart piloting system for the drone, which ensures both operational safety and high-quality visual inspection. The acquired data are securely managed by TÜV SÜD's inspection platform, which automatically masks any private information to protect your privacy.

The AI-based solution assists professional engineers in delivering detailed, accurate and compliant inspection reports. The software constructs a 3D model of the building façade, which helps to better understand the building structure and automatically locate the detected defects on the building.

The TÜV SÜD Drone Façade Inspection application provides access to all the data, report findings, and 3D model at any time. Repairs and follow-ups can be seamlessly managed through the platform to improve efficiency and save costs.

29.4.1 Solution Sketch

Principle stakeholders for operations of the drone in the field include the drone pilot, safety officer, and domain expert. Professional engineers are supporting in the backend. Customer stakeholders include building owners, facility managers, and regulators.

The drone is equipped with a number of sensors to support both flight operations and building façade scanning. These flight support sensors include IMU, UWB, Lidar and stereo cameras. The drone carries thermal sensors and a visual camera as the main payload for building façade data capture. On the drone, AI is mainly used for drone positioning, collision avoidance and path planning. This is supported by a smart controller device used by the drone pilot on the ground.

A number of backend applications support the management, processing, analysis and visualization of the captured data. Domain experts and professional engineers can add their domain expertise as well.

29.4.2   Drone Control

A key feature of the solution is advanced drone control, which provides semi-automated path control for scanning the building surface, supporting complex urban environments. Multimodal sensor fusion is used for navigation. Autopath planning supports inspection and obstacle avoidance and operational safety of the drone and ensures high-quality image capture for visual inspection.

To support this, the drone carries a miniature, high-performance Inertial Measurement Unit (IMU) and Attitude Heading Reference System (AHRS). The Lidar sensor provides stereo data for dense short range on path obstacle detection (30 m). The system also has two stereo cameras for sparse long-range obstacle detection (120 m).

29.4.3   Drone Data Analysis: Façade Inspection

Another key application of AI is drone data analysis, which is used for creating façade inspection reports. First, the raw façade data are preprocessed, e.g., anonymizing the captured data. Second, an AI-enhanced image analysis tool is applied to visual and thermal data. Finally, the meta-data are analyzed, utilizing AI to identify individual façade elements, different types of defects, and even detailed defect attributes.

29.5   Expert Opinion

The following discussion will provide insights into the TÜV SÜD Drone-based Building Façade Inspection project from Marc Grosskopf (Business Unit Manager, Building Lifecycle Services, TÜV SÜD, Germany) and Martin Saerbeck (CTO Digital Services, TÜV SÜD, Singapore).

Dirk Slama: *Marc, what were — or are — some of the biggest challenges in this project?*

Marc Grosskopf: *Only opportunities, no challenges! However, all kidding aside: of course this is an iterative process, from the initial pilots to the global roll-out which we are currently preparing. In the early stages, challenges tend to be more common on the technology and sourcing side. Then, you are quickly getting into regulatory aspects, customer acceptance, data quality, internal acceptance and processes, regional differences, etc. So it is never getting boring.*

DS: *Martin, from the CTO perspective, what were some of the initial challenges?*

Martin Saerbeck: *On the technology side, we have two main aspects: Drone-based image capturing and the data platform. For drones, it is very much about striking a good balance between cost, flight capabilities, and the quality of the sensors, and of course establishing a supply chain that can support us globally. For the data platform, we need to be able to support stakeholders with different backgrounds, roles and responsibilities. The user interface must be intuitive even if backend AI algorithms can be quite complex.*

MG: *Yes, do not forget that we have quite a complex constituency — drone operators, data scientists, domain experts, customers, and so on — all need to be supported by the central Façade Inspection Platform.*

DS: *Let's start by looking at the drones and drone operations...*

MS: *Of course you need to get the initial platform setup correctly. There are many powerful drone platforms available, but we need to adapt them to our needs, and not the other way around. One example is implementing automated flight path control to ensure façade coverage and high quality images. But perhaps the greatest challenge is keeping up with the constant flux of technology and changing regulatory requirements in different regions. Take, just as an example, free-flying vs. tethered (i.e., cable-bound drones). There are many different opinions on what should happen if the tether fails: Are we allowed to automatically switch over to the drone battery for safe landing or not? How much time do we have until we need to trigger an emergency routine? What exactly constitutes a tether failure? The list goes on. For us, it is important to be directly involved in standardization committee work, both locally and globally.*

MG: *Technical people tend to focus on the "sexy" stuff first: AI, automation, image analysis, and so on. However, we also need to look at drone maintenance, firmware updates and battery management. Of course, on-site support such as system setup, traffic management, etc. At the end of the day, this process needs to be so efficient and effective that the overall process is cheaper than the manual process. We need to ensure that we have enough in-house knowledge before we can source this regionally. We cannot take any shortcuts because we need a solid foundation and have to avoid building up technical debt because of cost-driven supplier situations.*

DS: *Let's talk about the backend platform. What does this look like?*

MG: *It depends on who you talk to. For drone pilots and on-site staff, the platform mainly needs to support the management of image uploads. For domain experts, we need an efficient way of reviewing and labeling the image material. This process is now increasingly supported by AI. Finally, we have end customers who access the platform to obtain the final results and reports. As an added challenge, they want to use the platform to monitor and manage building defects in the current project and for future comparison of quality development."*

DS: *Does this mean the platform is not magically smart and fully automates the inspection process from the start via image analysis?*

MG: *It gives us a fully digital process from the beginning, since we now have a process for efficiently capturing and managing the image data. This is already an important step. We are now gradually using our huge network of building façade domain experts to label relevant data and then use this to train the system. This means over time we get more and more automation. Initially, by prefiltering huge amounts of data, domain experts only have to review relevant image data. So this leads to more automated classification.*

DS: *How does this look like?*

MS: *Based on the labeled data from domain experts, our data scientists are accessing the platform via standard developer tools to build a library of defect detection algorithms. These algorithms vary depending on the defect type and the façade materials. For example, cracks need different detection algorithms than spallings; glass façades are different than metal or concrete.*

DS: *When can you retrain, and when do you have to develop new algorithms?*

MS: *It depends. For example, for cracks in different concrete types, we can use transfer learning to a certain extent. However, detecting and evaluating cracks in glass requires models that we essentially train from scratch.*

DS: *What about privacy?*

MS: *This is a very important point. Especially if the drone is likely to inadvertently capture people throughout the scanning process (e.g., standing behind windows), we need to automatically identify and anonymize this. Privacy preservation is key. We spent considerable effort on this portion.*

DS: *So how are your scaling this up for global roll-out?*

MG: *First, we have to ensure regional support for the drone service. This means dealing with local regulations, finding local service partners, suppliers and so on. Then, we have to ensure that our processes can be easily replicated: how do you execute a drone-based building scan, how are our domain experts working with the data, how can our central competence center in Singapore best support the regions with reusable fault detection algorithms, and how do we best onboard and support our customers in the regions?*

DS: *Your current focus is on building façades. Can you apply your lessons learned also to other use cases?*

MS: *Sure. Let us take, for example, building construction progress monitoring. There are many similarities here. This is an area where we are following a similar approach, together with our partner Contillio, which is focusing on Lidar and AI for analyzing the construction progress and mapping this back to the original BIM models. Of course you can also take a similar approach to inspection of power plants, bridges, solar panels, etc. A lot is happening, but we have to take it step by step!*

DS: *Thank you, Marc and Martin!*

Chapter 30
Predictive Maintenance for Hydraulic Components (Bosch Rexroth)

Dirk Slama

Fig. 30.1 ODiN Case Study

Predictive Maintenance has long been the holy grail of the IoT. However, experience has also shown that successfully implementing predictive maintenance for industrial use cases is harder than one might think, from finding a sustainable business model to actually delivering the technical implementation. This case study provides an account of a successful predictive maintenance implementation for hydraulic systems from the perspective of Bosch Rexroth, a leading supplier in this field (Fig. 30.1).

30.1 Hydraulic Systems

A hydraulic system uses pressurized fluids (usually mineral oil) to drive actuators to produce linear or rotational movements. Example use cases include hydraulic excavators, hydraulic presses, mining conveyor belts, shredders, hydraulic lifts, etc.

D. Slama (✉)
Ferdinand Steinbeis Institute, Berlin, Germany
e-mail: dirk.slama@bosch.com

Hydraulic components include cylinders and motors (to produce linear or rotational movements) and hydraulic power units to supply pressurized fluid to actuators. These consist of pumps, coolers, tank, etc. Valves are used to control the fluid flow and pressure. Hydraulic oil is not only used for power transmission but also serves as a lubricant and cooling fluid.

Benefits of hydraulic systems include:

- Simple generation of high forces (> 1x106 N) and torques (> 1x106 Nm)
- High power density
- Accurate control of high forces
- Robustness
- Simple, cheap and fast overload protection (Pressure Control Valve)

Hydraulic equipment vendors such as Bosch Rexroth supply machine builders with hydraulic components and systems either directly or via sales partners. Machine builders utilize hydraulic equipment to build industrial machinery, e.g., a hydraulic press, a plastic injection molding machine, or a conveyor belt for heavy loads. This machinery is operated by different types of operators. The hydraulic equipment vendor would usually also offer these operators different services, including spare parts, field service, and repairs. Without predictive maintenance, these services would naturally be reactive, i.e., only triggered after a problem with the hydraulic equipment in the field. This can lead to significant production outages. For example, the outage of a hydraulic component powering a conveyor belt at a mining site could lead to a shutdown of the entire mining operation (Fig. 30.2).

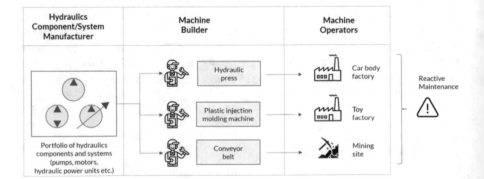

Fig. 30.2 Current Situation

30.2 Typical Problem Scenarios

What are the typical issues with hydraulic systems and components? A main reason for wear and breakdowns is contaminated hydraulic oil (by contaminants such as particles and water). This can lead to wear, which in turn can lead to reduced efficiency (e.g., increased volumetric losses in pumps, external leakage in cylinders due to worn seals) or malfunction (e.g., blocked valve spool). The result is lower efficiency or malfunction and increased breakdown probability. Breakdowns can be expensive: while the exact costs are usually use case specific, target customers for Predictive Maintenance typically have downtime costs exceeding 10.000 €/hour (Fig. 30.3).

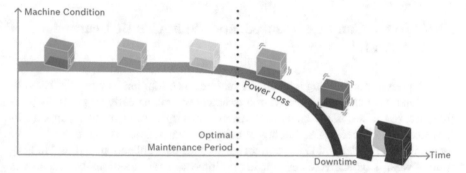

Fig. 30.3 Hydraulic systems: typical problem scenarios

Often, downtime is reduced by built-in redundancy. However, this cannot fully guarantee availability: cylinders are typically not redundant, a pump breakdown can contaminate the hydraulic fluid and cause other damage, such as valve malfunction due to contamination by particles. Cleaning the hydraulic fluid after a breakdown and replacing all damaged components can be very time consuming (this may take weeks). Large, expensive machines often do not have a replacement machine to continue production after a sudden breakdown. With no advance warning, diagnosis of the causes and decisions on necessary maintenance measures can take a long time. Existing fail-safes in the machine are built to shut down the machine after a catastrophic failure and protect the operators and the environment but rarely contain advance warning features.

30.3   Predictive Maintenance: Issues and Solutions

A key problem for building a predictive maintenance solution for hydraulic components is the complex, individual machine behavior:

- Many different products are produced with the same machine
- Hydraulics are usually only a small part of the whole machine
- Upgrades/changes to the machine after years of operation, e.g., new cooler
- Environmental effects: e.g., temperature, vibration
- Individual changes applied by machine operator

The result is that in most situations, there are initially insufficient data for building an end-to-end AI solution. This is why the Rexroth team has taken an approach where AI-based anomaly detection is used to find interesting data patterns. This is combined with human experts to diagnose the anomaly and subsequently make customer-specific maintenance recommendations.

30.4 What Can Be Measured, and What Can Be Learned from It?

A key question for building a predictive maintenance solution is: what can be measured, and what can be learned from it to detect wear at an early stage? In hydraulics, wear is a key issue. However, wear is very difficult to measure directly in practice. Wear processes and component/system functions must be understood in detail in order to determine the correct sensors for data collection. Indirect indication of wear is typically achieved using multiple sensors. Additionally, sensors for measuring the operating point of the components are required since many values, such as leakage and vibration, are operating point dependent. Commercially available sensors are used to reduce costs.

A good example is the external leakage on pumps and motors: Flow meters for leakage flow measurement, operating point: Pressure, speed, displacement, temperature. Another example is cavitation on pumps (suction flow of the pump is lower than vapour pressure due to contamination, excess speed, dissolved air in the hydraulic fluid, etc. Oil vapour bubbles are imploded during the transition to the high-pressure side and cause wear when this happens close to metal parts (e.g., distributor plate). Structure borne sound measurements with accelerometers are used to detect changes in the frequency spectrum of the structure borne sound. The operating point (pressure, speed and displacement) also has to be included.

30.4.1 Why Not Simple Rules-Based Analysis?

The next question is how to analyse this. Does it have to be AI, or could a simpler, rule-based or analytical model be applied? The problem with these approaches is complexity. While the hydraulic components are standardized, this does not apply to the machines built using them. Consequently, this would require new rules for each machine individually or models to be created and model parameters to be tweaked for each application, meaning a very high individual effort per customer. Furthermore, machine operation (e.g., dynamic operating points, variable environmental effects, changes in production, retrofits and modifications to the machine, etc.) would make the rules very complex and error prone, resulting in false alarms.

30.4.2   Why ML-Based Anomaly Detection, But Not Prescriptive Analytics/Automated Recommendations?

Because of the high complexity and missing labeled failure data of the individual customer environments, it has proven not to be feasible to apply an end-to-end AI approach, e.g., using deep learning with nonanalytic feature extraction using CNNs (Convolutional Neural Networks).

Consequently, the solution chosen by the Rexroth team is based on "classical" ML, using feature extraction (using domain-specific methods) and unsupervised Learning. Complex dependencies between features are solved by ML. The result is a working anomaly detection, but potentially with many possible causes.

This means that in addition to automated anomaly detection, a human expert is required for failure diagnosis and maintenance recommendation due to individual applications. It is also possible that changes were made to the machine, which cannot be measured with sensors (e.g., new cooling water supply) or that machine operators have changed the settings or are producing different products on the same machine (Fig. 30.4).

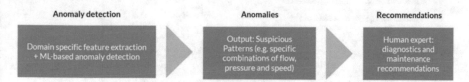

Fig. 30.4 ML Approach for Hydraulics Maintenance Predictions

The resulting approach is a two-step analysis process:

1. Machine Learning-based anomaly detection: Classic domain knowledge-based feature extraction + Machine Learning. Algorithm scans the data for interesting patterns. Output metrics, e.g., system behavior, are calculated and visualized on a GUI for human experts. Dashboards provide a quick overview of machine behavior.
2. Human experts diagnose suspicious data patterns based on general domain experience and application/customer-specific know-how. Sometimes it is necessary to ask the customer for further details (e.g., if mechanical modifications have been made to the machine or settings/parameters have been changed). This manual work is necessary as not everything can be captured in the data.

30.5   The ODiN Solution Offering

Based on the capabilities but also the limitations of the ML-based approach, the Bosch Rexroth team decided to build the ODiN solution, which is a predictive maintenance service consisting of:

- Application of a specific sensor package to be retrofitted into the customer machine
- Data acquisition unit and IoT gateway for cloud connectivity
- AI pipeline in the cloud
- Personal service support in case of anomalies and quarterly status reports
- Optional additional services, e.g., spare parts management, field service, repairs

The maintenance contract is signed with the machine operator. Maintenance can be carried out by Rexroth, a Rexroth service partner, the customer or a maintenance contractor. Maintenance contract templates are country unit specific and may be customer specific. The contract always contains an appendix detailing data use (Fig. 30.5).

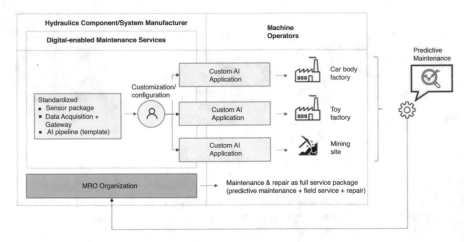

Fig. 30.5 Predictive maintenance solution

30.5.1   Customer Offering

The offered solution is a one stop shop for predictive maintenance covering everything from application-specific engineering to maintenance recommendations and data transmission as well as a secure operation of the data platform. A monthly fee is charged for the service, and parts of the contract are charged as a one-time payment (e.g., installation of data acquisition) (Fig. 30.6).

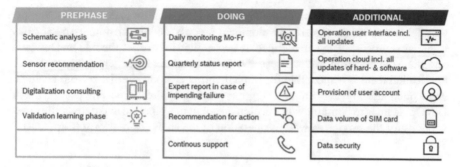

Fig. 30.6 Predictive maintenance offering: as-a-service

30.5.2   Lifecycle Perspective

The target customers are machine operators with high downtime costs. These machines are typically already in the field and have been operating for many years. Existing sensors do not provide enough data for a reliable diagnosis. Therefore, a retrofit sensor package and data acquisition unit must be installed onsite. After commissioning, data are sent to the cloud and stored on Bosch servers to be analyzed. ML-based anomaly detection provides insights into general machine behavior, and a human expert will offer maintenance recommendations to customers if required. Additionally, experience from field data is fed back to the continuous development of the ODiN platform and analytics solution (Fig. 30.7).

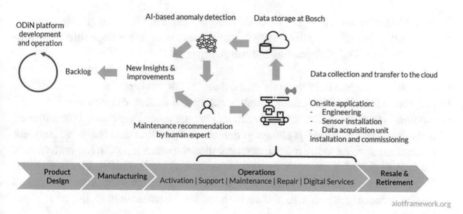

Fig. 30.7 Lifecycle Perspective

30.5.3   Customizing the ML Solution

Because of the high level of heterogeneity found at customer sites, efficient customization of the solution is important. The approach taken will be explained in the following.

Development and Customization Processes The solution is developed using two parallel processes: the generic development process, and the customer-specific customization process. They are defined as two individual cycles: the AI DevOps cycle and the AI application cycle. These two are carried out by separate teams. The AI application team is responsible for implementing customer projects from customer acquisition all the way to monitoring the running applications. The task of the AI DevOps team is to continuously develop the analytics pipeline and deliver improved versions for the service as well as operation of the analytics platform (Fig. 30.8).

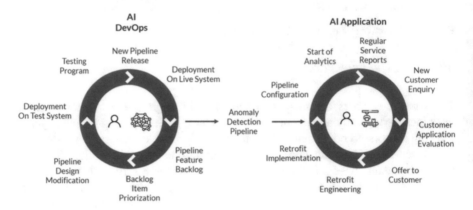

Fig. 30.8 Customization Process

Technical Approach A single, generic, analytics pipeline for anomaly detection is used for all applications. This enables scaling, as no customer-specific programming is required. The pipeline has the following steps:

- Data export: export data from the big data store for analysis
- Preprocessing: domain-specific preprocessing and feature extraction
- Anomaly detection: automated Machine Learning model generation for anomaly detection. The first model is always generated with the first data batch. Subsequent batches are applied to the model, and new model generation with the current data batch is triggered if the error exceeds a predefined limit. This results in a model library with each model describing a specific machine behavior. These behaviors can be manually labeled to create metrics for visualization in the next pipeline step
- Post-processing: generation of metrics for visualization and monitoring of applications
- Publishing: Calculated metrics and logs are published to kafka

The pipeline is configured for each application via a JSON configuration file, which contains sections for each pipeline step. This enables application-specific analyses without customized programming work (Fig. 30.9).

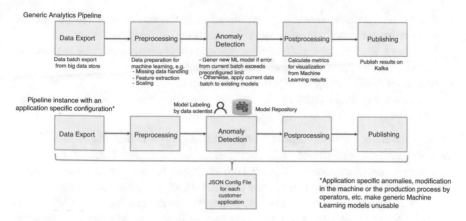

Fig. 30.9 Customization Details: Technical Approach

30.6   Customer Example

Lafarge Holcim is a global supplier of cement and aggregates (crushed stone, gravel and sand), as well as ready-mix concrete and asphalt. In their cement manufacturing facility in Bulacan, Philippines, Lafarge Holcim's challenge was to monitor the key indicators of the hydraulically operated clinker cooler in order to detect possible failures in good time.

A clinker cooler is an essential component of cement production. If it stops working, the entire production must stop within five minutes. Therefore, the sensors on the hydraulic system were installed in such a way that they send the essential physical quantities to the ODiN platform for analysis. Employees of Lafarge Holcim are receiving regular reports about the system behavior of their machine. A local service partner interprets the information provided by the ODiN system and gives recommendations for action to the maintenance technicians on site.

Feedback from Lafarge Holcim: *Now we are able to learn much more about our own equipment than we did before. We can predict, we can see the health of the machine. We will install it to other hydraulic units within our business. With ODiN, we can be more proactive in capturing all the equipment data, making better decisions.*

30.7   Summary and Lessons Learned

The following provides a summary and key lessons learned:

- Retrofitting sensors are necessary for required data quality; specialized data acquisition needed

- Application specific anomalies, modification in the machine or the production process by operators, etc. cause generic Machine Learning models to fail
- Building a working generic analytics pipeline for anomaly detection is possible, however, with application-specific configuration
- Manual model labeling by experts is necessary
- A human expert is required for failure diagnosis due to complex machine behavior

Perhaps the most important lesson learned in this project is that due to insufficient data for generic, end-to-end ML solutions, a low-cost solution for Predictive Maintenance of heterogeneous industrial environments is not realistic. Consequently, the team decided to offer a full service contract together with personalized support to maximize customer value. The combination of human expertise with ML-based anomaly detection enables a reliable and efficient Predictive Maintenance solution for the customer, helping to significantly reduce downtime and improve OEE (Overall Equipment Effectiveness).

Chapter 31
BaseABC: Addressing the AIoT Long Tail in I4.0 (Bürkert)

Nikolai Hlubek

This case study highlights how Bürkert Fluid Control Systems, a successful I4.0 SME, masters the AIoT long tail by applying the BaseABC method. The case study was authored by Dr. Nikolai Hlubek, who works as a Senior Data Scientist at Bürkert Fluid Control Systems and develops new data-driven products. The author has a PhD in physics and has been using data science for more than 15 years to tackle various topics. As an example, before joining Bürkert Fluid Control Systems, he developed a real-time ionospheric monitoring service for the German Aerospace Center.

31.1 Introduction to Bürkert

Bürkert Fluid Control Systems develops and manufactures modules and systems for the analysis and control of fluids and gases. Examples of typical products include large process valves for the food and beverage industry, small electrodynamic valves for pharmaceutical applications, mass flow controllers, sensors for contactless flow measurement based on surface acoustic waves, and sensors to measure the water quality of drinking water. Bürkert is a 100% family-owned company that employs approximately 3000 people, has a consolidated turnover of ~560 M€, is headquartered in Ingelfingen (Germany), and has locations in 36 countries worldwide (Fig. 31.1).

N. Hlubek (✉)
Development Engineer, Dresden, Sachsen, Deutschland
e-mail: nikolai.hlubek@bukert.com

© The Author(s) 2023
D. Slama et al. (eds.), *The Digital Playbook*,
https://doi.org/10.1007/978-3-030-88221-1_31

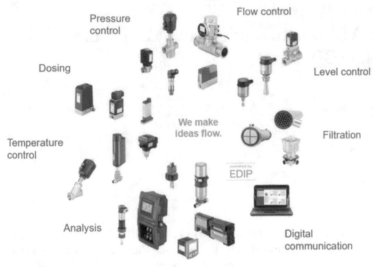

Fig. 31.1 Introduction to Bürkert

Bürkert products have a moderate level of complexity, which means they can be developed in small project teams of usually less than 10 people over the course of one year. However, Bürkert has a very large portfolio of such products. This portfolio structure places Bürkert at the long tail of AIoT, where it has to manage product variants in a very efficient way. Therefore, product development at Bürkert is truly a good testbed for any AIoT long-tail development process, as the entire process is repeated many times in a relatively short period of time, due to the parallel development of many small products and the relatively short development time. The approach is now a well-documented best practice at Bürkert, as will be explained in the following.

31.2 BaseABC

31.2.1 BaseABC

Figure 31.2 illustrates the BaseABC method that Bürkert uses as a workflow for its data science projects. The workflow does not distinguish between information visualization, algorithm development or machine learning, as the fundamental steps are the same in all cases. The workflow is highly iterative. The workflow is restarted anytime a new insight arises from a step that has implications for a previous step. For a new idea, the initial completion of the workflow will deliver a technological demonstrator. Successive iterations will build on this and deliver a prototype, a minimal viable product and finally a saleable product. Due to the iterative nature of the workflow, it can be stopped anytime with the minimum amount of time and money invested, should it become clear that the data science project would not be able to be transitioned into a sustainable business model.

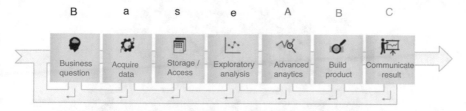

Fig. 31.2   BaseABC method for I4.0 data science projects

The workflow starts with a *business question*. Such a business question could be a pain point for a customer, which can be mitigated by a new product that uses additional data. It could also be an idea for improving an existing product by using data science tools. It could also be an enhancement of an existing data-driven product in the field in the form of continuous learning. The next step is to define a data collection strategy and to implement that strategy, i.e., *acquire data*. We check whether the data are already available, whether we need to generate it, what quality we require and what tools are necessary to collect the data. Then we collect the data using the defined strategy. Then, *storage and access* to the data must be considered. We store the acquired data in such a way that access in the following steps is easy, computationally efficient and future proof. An *exploratory analysis* of the acquired data follows. A domain expert checks the data for consistency and completeness. He or she investigates isolated issues with some of the datasets, if any, and decides if these datasets need to be acquired again.

If all these prerequisites are in place, a data scientist begins the work in the *advanced analytics* step. He or she tries to find a solution to the business question with the help of data science tools. Once the data scientist has found a solution, we *build* a product as quickly as possible. The first iteration is, of course, a technology demonstrator. A prototype, a minimal viable product and the actual saleable product follow in consecutive iterations. When the product is ready, we *communicate* the result. In an early iteration, this is an internal review of the technology demonstrator. In later iterations, a pilot customer obtains a prototype for testing and feedback. Finally, this step will mark the introduction of a new data-driven product.

An example of a product that we developed at Bürkert using this workflow is a diagnostic application for a solenoid valve (Bürkert type 6724). These valves are small electrodynamic valves for different media, such as air and water, and can be used in dosing applications as shown in Fig. 31.3. In a typical variant, they are held closed by a spring and opened for as long as a current is applied. The behavior of the current during the opening of the valve - the so-called inrush current - can be used for diagnostics. The counter-electromotive force is a part of the inrush current and is proportional to the actuator movement. Therefore, it is possible to assess the dynamics of the actuator by a cheap current measurement using the actuator of the valve as a sensor. In particular, it is possible to check if the valve truly opened or if some blockage occurred without the need for any external sensors.

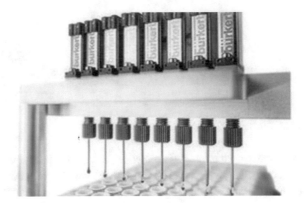

Fig. 31.3 Bürkert valve type 6724 in a typical dosing application

The graph in Fig. 31.4 shows an example of such current curves. It shows two curves where the valve fully opens (100% stroke) and a curve where the valve only partially opens (50% stroke), i.e. where the fluidic channel is blocked. We can see that all curves are quite different from each other. In particular, the two good state curves (100% stroke) are different in shape. The reason for this is that the counter-electromotive force is only a part of the inrush current, and its exact shape depends on many internal parameters (diaphragm type, coil type, …) and external parameters (temperature, pressure, …). A possibility to estimate the movement of the actuator is by using a curve shape analysis of the inrush current for a data set that contains measurements with suitable combinations of all parameters. This is a standard machine learning task. We will use this example in the following to explain the workflow in detail.

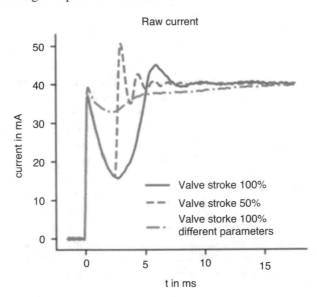

Fig. 31.4 Raw current measurement for the first few milliseconds when switching an electromagnetic valve (Bürkert type 6724). The blue solid and gray dashed-dotted lines show the switching current for a working valve. The orange dashed line shows the switching current for a blocked valve

31.2.2   Pipeline

Bürkert envisions the workflow as a pipeline. The complete workflow is started over every time new insights arise from a step that requires an adjustment of previous steps. We then ask for each step if we need to adjust it, based on the new insights. For this to be feasible, the implementation of the workflow must be automated as much as possible. Any manual task required during the workflow represents a pain point that makes it unlikely that the workflow will actually work as a pipeline.

In other words, the time, effort and cost of going through the workflow repeatedly must be as low as possible. Otherwise, people will just find excuses and either not do necessary adjustments or find workarounds that introduce a long-term maintenance burden. Of course, not every step can be automated, e.g., rechecking the business question. However, most of the time, this is also not necessary and we are good enough if the most time-consuming steps acquire data, storage and access and advanced analytics of the final solution are mostly automated. In the case of the business question it is usually sufficient to check if the new insight affects the business question, which is most often not the case. In the following, we illustrate this idea of a pipeline with examples that highlight how a step can lead to new insights that require an adjustment of the previous steps:

After coming up with a business question, we need to consider how to acquire the necessary data. It may turn out that it is not possible to obtain such data. For example, a client might not be willing to share the required data because it contains their trade secrets. Without an update to the business question, which must now include a way to protect the client's secrets, the workflow cannot continue.

A domain expert that explores the acquired data might notice that the data have certain issues. The measured data could show drift due to changes in ambient temperature. In this case, the data must be measured again under controlled environmental conditions, as any follow-up analysis of bad data is scientifically unsound and a waste of time for the data scientist.

After we built a prototype and provided it to a pilot customer, they may request an additional feature that requires additional data and an update to the solution implemented in the advanced analytics step. The customer may want to operate the device in a different temperature range. Thus, we need to acquire additional data in this temperature range and adopt our solution accordingly, hence rerunning the pipeline.

All these examples show that the data science workflow is repeated many times. This justifies the requirement that the workflow must be as automated as possible so that simple adjustments to the product stay simple in the implementation. We want to react quickly to customer requests without working more.

Up until a minimal viable product is ready, each step and iteration is a stop criterion for the whole project. Therefore, we strive to keep the number of complex tools at a minimum in order to keep the initial costs and the maintenance burden small.

For our diagnostics project of a solenoid valve at Bürkert, we automated the data acquisition by developing an automated measurement setup using LabView. We made the explorative analysis easy for the domain expert by providing a tool for quick visualization of current curves. We automated the advanced analytics step so that any stored data would automatically be detected and would go into the classifier. The classifier could then be deployed to the technological demonstrator and later to the prototype automatically.

31.3 Details for Individual Steps

In the following, we will explain the individual steps of the data science pipeline in more detail.

31.3.1 Business Question

The business question is the starting point for the data science pipeline. A good business question has a well-defined idea and measurable objectives. It includes a basic idea for the acquisition of useful data.

Example Project The business question for our example project was as follows: Is it possible to build a classifier that reports successful valve opening if more than 90% flow is achieved and reports an error for any smaller flow by monitoring the inrush current only? The classifier should work for all conditions allowed by the valve datasheet. Data acquisition should be done by laboratory measurements. This question is visualized in Fig. 31.5. It shows the inputs and outputs of the classifier that is to be developed. The parameters are not input to the classifier but complications that make the classification more complex.

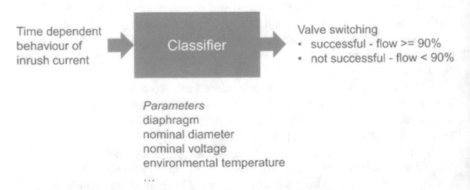

Fig. 31.5 Task of the data science project that we use as an example to illustrate BaseABC. The input to and required output of the classifier are shown. The parameters are not input to the classifier, but complications that make the classification more complex

31.3.2   Acquire Data

We define a strategy for the acquisition of the data and metadata. This means we check whether the data are already available, whether we need to generate it, what quality we require and what tools are necessary to collect the data. Then, we acquire the data and ensure accurate tracking of provenance.

Design of Experiment If we acquire the data through measurements, each experiment costs time and money. To optimize the number of required experiments for a maximum variance in the dependent variables, the design of the experimental method can be used.

Data Provenance Metadata will always be necessary to understand our data and to track how we obtained our data. When we acquire the data, we need to document our setup and the process of acquisition. Both documents will serve as an explanation for our data. For the setup, we need to document which devices and tools we used, how we connected the devices and so on. For the process, we need to document how we acquired the data, in which order we acquired the data, when we acquired the data, which environmental conditions we used and so on. In short, we need to use good laboratory practice.

We save the resulting data sets and corresponding metadata in a structured form. Data sets and metadata must be automatically readable and parsable.

Labeling We label data during the data acquisition step if labeling is necessary and if it is possible. Labeling at a later stage always introduces the danger of hindsight bias.

Example Project Valve experts at Bürkert defined which range of parameters and fault states are relevant. We modified a valve so we could simulate fault states. We automated the acquisition of the current curves. In total, we acquired approximately 50,000 current curves for 12 parameters. Our automated setup stores the data (current curves) and metadata (internal and external parameters of the measurement) in the same measurement file. We generate one file of approximately 3 MB per measurement, which netted us with approximately 150 GB of overall data for the project.

31.3.3   Storage/Access

The best way for storage and access is to use the principles for fair data [29]. They state that the data should be findable, accessible, interoperable and reusable. Any storage solution that will fulfill these requirements is sufficient.

If a data archive system is not used and the data are stored on a fileserver, the storage should be set to read-only after some initial time that is reserved for quick fixes. This guarantees that any following data analysis will be reproducible because it uses the same data.

The data should be stored in a form that is easy to handle. For many projects, this means using a database is unnecessary. In our experience, for everything up to a few 100 GB, storage on a file server and lazy loading were computationally efficient, easy to handle and did not have the maintenance burden that a database would have introduced.

Example Project In our example project, we stored the files of our measurements on a file server. We compiled the metadata in tabular form with links to the data files. Using these links, we employed a strategy of lazy loading, i.e., loading the measurement data of the relevant curve into main memory only when required. Since we used a server-based approach for the data analysis, with the server in the same data center as the data, access to the data was fast. We profiled access times using our lazy loading strategy against using a NoSQL database. The results showed that the database was not faster.

31.3.4 Explanatory Analysis

Before any in-depth analysis of the data starts, we perform simple visualizations of the acquired data to obtain an overview of it. A domain expert uses these simple visualizations to check the data for consistency and investigates some isolated issues if such issues are present in the data.

Example Project We examined some of the current curves using simple plots of current versus time. For some data sets, a domain expert observed that the current curves had an inflection point at an unexpected position. Further analysis with the help of a data scientist revealed that all data sets with this feature belonged to a particular valve. Analysis of that valve revealed that it performs its function like the other valves but has a much higher friction due to manufacturing tolerances. This insight leads to an update of the data acquisition step with an additional parameter.

31.3.5 Advanced Analytics

Once the data and metadata are available and of good quality, a data scientist uses her or his tools of the trade during the advanced analytics step. A measurable objective exists in the form of the business question.

Data Visualization A first step of the analysis is usually to create a meaningful visualization of the data. During the exploratory analysis step, the domain expert was looking at individual datasets alone. In this step, the data scientist should design a visualization that encompasses all of the data. For example, this can be achieved by clustering algorithms such as principal component analysis for datasets that consist of sets of measurement curves.

Find Numerical Solution   Once the data scientist has gained an overview of the data, he or she searches for a suitable solution to the objective set by the business question. The solution can be in the form of an algorithm or a machine-learning model. It also does not matter whether the machine-learning model is a greedy model such as kNN, a shallow learning model such as a support vector machine, a random forest or gradient boosting, or a deep learning model. The data scientist has all information at hand to find the best solution to the business question. Best in this sense is the simplest solution that can solve the business question. If performance metrics are involved, usually speed of evaluation, model accuracy and reliability and required computational resources need to be balanced for an optimal solution.

Baseline Solution   The data scientist should always evaluate her or his solution against a simple baseline solution. This is required to prove that a more complex solution is necessary at all. By comparing the final solution to the baseline solution, the gain in efficiency can easily be shown.

Document Results and Failures   When data scientists try to find a solution, they will naturally encounter dead ends. Some methods might not work at all or might not work as expected. Machine learning methods are likely to find several solutions of varying quality. We document all these results and failures. All results are documented so that the best result can be selected. As stated in chapter- Find numerical solution, the best result is not necessarily the one with the highest accuracy, as other considerations such as computational efficiency or reliability can be of higher priority. All failures and dead ends should also be documented. There are several reasons for this. First, the failure might be due to some mistake on the part of the data scientist and a future evaluation could correct this mistake. Second, by documenting that a method did not work this method does not need to be tested in similar future analysis tasks. This prevents someone from making the same mistakes again. Third, if conditions change, this method could suddenly work. If it has been documented under which assumptions the method failed, it can be assessed if it is worth trying the method again.

Archive Data Analysis and Tools   The result of the advanced analytics step will be a solution to the business question. Either this solution will be incorporated in a product or a business decision will be based on that solution. To maintain a product or justify a business decision it must be possible to reproduce the exact solution at a later date. This is only possible if the data are preserved, the analytical work of the data scientist is preserved and the tools that the data scientist used are preserved. Data preservation is a prerequisite in the step Storage/Access. The analytical work of the data scientist must be stored in an archive system such as Subversion or Git. The tools must be stored in exactly the same version that the data scientist used. For programming languages, the libraries used must also be taken into account.

Analytics Expert Review The data analysis should be reviewed by another analytics expert. This can be done either in a tool-assisted fashion using a code review tool such as Upsource or Phabricator on each increment of work or by a walkthrough of the analysis after a part of the analysis has been completed.

Example Project To obtain an initial visualization of the overall data we used principal component analysis and colored the data according to the parameters we had defined. Figure 31.6 shows such a plot for the first two principal components colored by the parameter temperature. The complete visualization would encompass the first four principal components and individual plots for all parameters.

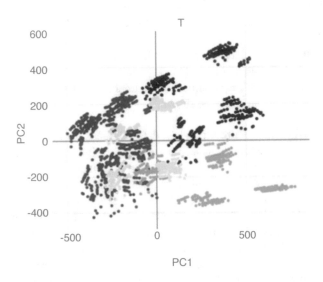

Fig. 31.6 Principal component analysis of the current curves. Shown are the first two principal components with the temperature show by color

Then, we developed our classifier using Jupyter notebooks [30]. These notebooks have the advantage that they can combine code for data analysis and explanatory text can include interactive figures and contain the results of the data analysis. They are a powerful tool for handling steps 3.5.1 through 3.5.5 in one view. The notebooks run on and are stored on a server. The source code in the notebooks is automatically exported when changes are made to the notebook. We archive the notebooks and the source code. The archive system is connected to a code review tool. We review each increment of work.

To work with reproducible tools, we use a virtual environment with fixed versions of the Python programming language and its libraries. This virtual environment is registered to the Jupyter notebook server and can be selected by a notebook for an analysis task. When the use of a new or updated library is required, we create a new virtual environment with the required version of the Python programming language and libraries, fix the versions and link it to the Jupyter notebook server. As a result, the existing notebooks use the old virtual environment and keep functioning. For new notebooks, the new or old environment can be selected.

Figure 31.7 shows our resulting classifier tested against an independent ground truth. Each point is a measurement with different parameters. For some measurements, a valve blockage was simulated. The classifier divides the measurements into good (circles, above the dashed line) and faulty (crosses, below the dashed line) states. Coded by color is the actual stroke, which was independently obtained by a laser distance measurement. This ground truth shows that the classifier did classify all the tested cases correctly. This figure and the underlying classifier are the work products of the data scientist from the advanced analytics step. This is the result, which we hand over to the next step.

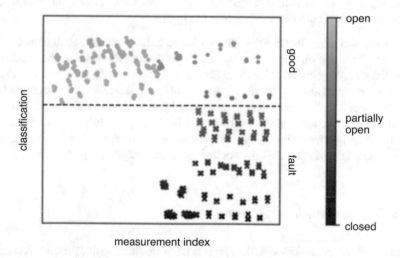

Fig. 31.7 Classification of valve stroke into good (circles, above the dashed line) and faulty states (crosses below the dashed line) based on the shape of the current curve. Coded by color is the actual stroke, which was independently acquired by a laser distance measurement

31.3.6 Build Product

When the data scientist finds a solution to the business question, he or she should isolate the required method to solve the question. The method can be an algorithm or a machine-learning model.

Deployment To ensure that the workflow functions as a pipeline the final solution of the data scientist should be in a way that it updates itself when new data are available. Additionally, it should be possible to integrate the solution into the product that is built in this step in an automated fashion.

This product will usually be a technology demonstrator in the first iteration of this workflow. In later iterations of this workflow, a prototype and minimal viable product might follow.

Domain Expert Review The technological demonstrator, prototype or product should always be checked by a domain expert. This review should be a black box review of the data scientist's solution. The domain expert should only evaluate the effectiveness of the demonstrator with regard to the business question.

Example Project The figure- Classification of valve stroke into good and faulty states above was the direct result of the advanced analytics step and used as initial technology demonstrator. In its first iterations, it contained a few measurements that were not classified correctly. The domain experts determined that the measurements were for valve states that were not allowed by the datasheet of the valve. Thus, after removing these faulty measurements from the dataset and redoing the analysis, we obtained the ideal classifier as shown in the figure.

Since this is a new technology, we decided to build a technology demonstrator in hardware. This demonstrator consists of a valve where the stroke can be reduced by an added screw to simulate blocking. The valve is connected to compressed air, and a flow sensor measures the resulting flow to obtain a reference measurement to be used as ground truth. The current is measured by a microcontroller, which also performs the classification. It shows the result by a simple LED. This technological demonstrator is important because it shows the effect of the technology without the mathematical details that are only accessible to an analytics expert.

31.3.7 Communicate Result

Once a technological demonstrator, prototype or minimal viable product is ready, it should be presented to a larger audience. This has multiple purposes. Inside a company, it is necessary to demonstrate the effect of the new technology on an audience that is not familiar with the details and gather feedback. It might also inspire people to use the technology for different products. Outside of the company, it is necessary to inform the technology to find pilot customers for field validation.

Field Validation Technological demonstrators will usually be shown around internally or at exhibitions to gather some initial feedback on the technology. If a prototype is available, field validation is the next logical step. This will present a real-world usage for the prototype. It will give valuable insights that more often than not lead to necessary adjustments of earlier steps of the BaseABC.

Business Case Review All the collected feedback should lead to an answer to the business question, whether the developed technology is able to solve the business question and whether the business question correctly addresses a customer issue. One might argue that the detailed identification of the customer issue should be earlier; however, real-world examples have proven time and time again that most customers can define their needs the best if they are given a prototype that deals with their issue. This is the main reason this workflow aims at automation and iteration toward an adoptable prototype as early as possible.

Example Project We patented the technology for the diagnostics of the inrush current. The initial business question spawned a number of follow-up projects with slightly different business questions, which all try to capitalize on the technology in one way or the other.

Pilot customers evaluated prototype circuit boards. Such a board has a microcontroller that measures the current and classifies valve switching. Customer feedback showed that these prototypes address customer needs.

The technology is ready for integration into systems, designed for specific customer applications, and a project to develop a standard product, which is done as another iteration of BaseABC, is on its way.

31.4 FAQs

31.4.1 Why Is There No Step for Continuous Training

Training a machine learning model on historical data and deploying it assumes that the new data will be similar to the historical data. Sometimes this is not the case, and the machine-learning model needs to be adapted to retain its accuracy and reliability. This concept is called continuous training.

Within the BaseABC workflow, this is not an extension of an existing workflow. In some cases – mostly when the data-driven product is cloud based - it is just a rerun of the existing pipeline. If the data-driven product involves an edge device and access to new data is difficult, we handle it as a separate business question that uses its own pipeline. This is justified because, for example, the data acquisition and storage usually differ drastically in such a case from the initial model training. Typically, the initial training is against a large data set and happens with some powerful infrastructure involved. Later, data collection and computationally less expensive model refinement occurred on the edge device.

31.4.2 Why Is There No Step for Monitoring an Existing Data-Driven Product

Similar to continuous training, the BaseABC workflow treats the monitoring of a data-driven product as an individual business question. It requires the acquisition of appropriate monitoring data – usually containing a ground truth, a storage location (edge device or cloud) and a suitable performance metric for monitoring the prediction quality that the data scientist has to find.

Appendix A: Editorial Board

TANJA RÜCKERT
(Chairwoman of the Editorial Board)
Group CDO, Robert Bosch GmbH

As chief digital officer of the Bosch group, Tanja is responsible for digital transformation and enabling scalable digital business models. Before this, she has been President of the Board of Management for Bosch Building Technologies since 2018. Before joining Bosch, Tanja was President of the IoT and Digital Supply Chain business unit at SAP SE. She holds a PhD in chemistry and has more than 20 years of experience in the software and building industry.

PRITH BANERJEE
Chief Technology Officer, ANSYS
Board member, Cray and CUBIC

Prith Banerjee is currently the Chief Technology Officer at ANSYS and Board member at Cray and CUBIC. Prior to that, he was EVP and CTO of Schneider Electric. Formerly, he was Managing Director of Global Technology R&D at Accenture. Earlier, he was EVP and CTO of ABB. Formerly, he was professor of ECE at the University of Illinois. In 2000, he founded AccelChip which was sold to Xilinx Inc.

Prof. JAN BOSCH
(Scientific Advisor)
Professor at Chalmers University
Director of the Software Center.

Executive, professor and consultant with more than 20 years' experience in large scale software R&D management and business. Extensive consulting experience with Fortune 500 companies and board member & angel investor in startups.

KEN FORSTER
Executive Director, Momenta Partners

© AIoT User Group 2023
D. Slama et al. (eds.), *The Digital Playbook*,
https://doi.org/10.1007/978-3-030-88221-1

Research Fellow at Ferdinand-Steinbeis-Institut

Dirk Slama is VP and Chief Alliance Officer at Bosch Software Innovations (SI). Bosch SI is spearheading the Internet of Things (IoT) activities of Bosch, the global manufacturing and services group. Dirk has over 20 years' experience in very large-scale distributed application projects and system integration, including SOA, BPM, M2M and most recently IoT. He is representing Bosch at the Industrial Internet Consortium and is active in the Industry 4.0 community. He holds an MBA from IMD Lausanne as well as a Diploma Degree in Computer Science from TU Berlin.

SEBASTIAN THRUN
President, Udacity
CEO, Kitty Hawk

Sebastian Thrun is an entrepreneur, educator, and computer scientist from Germany. He is CEO of Kitty Hawk Corporation, and chairman and co-founder of Udacity. Before that, he was a Google VP and Fellow, a Professor of Computer Science at Stanford University, and before that at Carnegie Mellon University. At Google, he founded Google X and Google's self-driving car team.

NIK WILLETTS
President and CEO, TM Forum

As President & CEO of TM Forum Nik Willetts is passionate about the digital revolution and the impact technology and telecommunications can have on business, cultural, societal and environmental issues. Prior to joining TM Forum, Nik ran a successful consulting business working with multi-national companies across the telecoms sector.

Appendix B: Expert Network

Core Contributors

DIRK SLAMA
(Editor-in-Chief)
EXPERT

Dirk Slama is VP and Chief Alliance Officer at Bosch Software Innovations (SI). Bosch SI is spearheading the Internet of Things (IoT) activities of Bosch, the global manufacturing and services group. Dirk has over 20 years' experience in very large-scale distributed application projects and system integration, including SOA, BPM, M2M and most recently IoT. He is representing Bosch at the Industrial Internet Consortium and is active in the Industry 4.0 community. He holds an MBA from IMD Lausanne as well as a Diploma Degree in Computer Science from TU Berlin.

EXPERTISE:
- AIoT
- Architecture
- Methodology

DAVID MONZEL, MM1
EXPERT

David Monzel is a senior consultant at mm1. During his studies of industrial engineering, he gained experience in the fields of connected and AI-supported manufacturing, project management, agile development and consulting through various internships at manufacturing and consulting companies. At mm1, he actively contributes with his expertise in the fields of AIoT, digital & agile transformation and strategy in large IoT and IT projects across various industries

EXPERTISE:
- Digital Transformation
- Hybrid Product Development

© AIoT User Group 2023
D. Slama et al. (eds.), *The Digital Playbook*,
https://doi.org/10.1007/978-3-030-88221-1

HEIKO LÖFFLER, MM1
EXPERT

Heiko Löffler is a consultant at mm1. During his studies of industrial engineering, he gained experience in the fields of smart connected products, industry 4.0, financial risk management and consulting through various internships at companies such as TRUMPF GmbH and SICK AG. Meanwhile, he is working as a consultant for mm1 in large IoT projects.

EXPERTISE:
- Industrial IoT
- Servitization (Equipment-as-a-Service)

CK VISHWAKARMA, ALLTHINGSCONNECTED
EXPERT

CK has an extensive international experience in strategy, business & operational excellence, program management, complex solutions in multi-techs, multi-vendor's scenarios. He is a hands-on team leader and has successfully initiated, led, and managed multimillion USD complex systems and solutions integration projects in his 17+ years of industry experience.

EXPERTISE:
- Digital Transformation
- Business Strategy

SEBASTIAN HELBECK, ROBERT BOSCH POWER TOOLS GMBH
EXPERT

As platform owner at Bosch Power Tools, Sebastian is responsible with the project team for the strategy and execution of the embedded drive train, IoT connectivity, field data and embedded AI. Beforehand he was Head of Engineering Bosch Connected Mobility Solutions since 2018. Sebastian studied electrical engineering, joined Bosch 2001 and was in charge of different international leadership positions.

EXPERTISE:
- Platform Owner Drive Train

MARTIN LUBISCH, BOSCH CENTER FOR ARTIFICIAL INTELLIGENCE
EXPERT

As an AI Consultant and Product Manager at the Bosch Center for Artificial Intelligence Martin Lubisch identifies solutions that help customers in their daily business and brings those to scale e.g., Vivalytic, a COVID-19 testing device enhanced with DL based optical inspection. Martin has co-founded a start-up and worked along the lines of Industry 4.0, IoT and AI in various locations such as Berlin, Stuttgart and Silicon Valley.

EXPERTISE:
- AIoT
- Business Development
- Product Management

ERIC SCHMIDT, BOSCH CENTER FOR ARTIFICIAL INTELLIGENCE
EXPERT

Eric Schmidt is an AI consultant and data scientist at the Bosch Center for Artificial Intelligence (BCAI). He initiates and implements data-driven solutions in various areas, ranging from engineering to manufacturing and supply chain management. Eric holds degrees in both computer science and business administration.

EXPERTISE:
- AIoT
- Machine Learning
- Business Development

HAAS PHILIPP, ROBERT BOSCH GMBH
EXPERT

Dr. Philipp Haas is a lawyer at Robert Bosch GmbH in the legal department. He heads the Expert Group for Digital and New Businesses. His field of activity for many years has included the drafting and negotiation of software license agreements.

EXPERTISE:
- Law

MARC GROSSKOPF, TÜV SÜD
EXPERT

Marc started at TÜV SÜD in 2006 in the structural engineering department focusing on qualification and technical design. He has vast experience in dealing with demanding construction projects in a wide variety of industries. Prior to his current role, Marc was the department head for the Real Estate & Infrastructure division in South Korea. From 2017 on Marc has been assigned as business unit manager, for TÜV SÜD global activities in the field of building lifecycle solutions focusing on digital service integration.

EXPERTISE:
- Construction Projects

MARTEN OBERPICHLER, BOSCH
EXPERT

Marten Oberpichler is a working student at Bosch - Central Department IoT and Digitalization. After his Bachelor of Science, he started his Master of Science in Industrial Engineering in 2018. Marten has over two years of work experience at Bosch.

EXPERTISE:
- AIoT

RALPH NELIUS, DEUTSCHE POST AG
EXPERT

Ralph Nelius loves to build great products in agile teams. After various stints as a software engineer, consultant and enterprise architect, he now works as a product owner at Deutsche Post on AI topics.

EXPERTISE:
- Product Development
- Analytics
- AI

DR. MARTIN SAERBECK, TÜV SÜD
EXPERT

In his role as CTO Digital Service at TÜV SÜD, Dr. Saerbeck oversees the technology roadmap and key implementation projects of digital testing services, including the use and assessment of AI. He has a long track record in academia and industry in the domains of smart sensor networks, robotics, and AI. After completing his PhD with Philips Research, Dr. Saerbeck started an interdisciplinary research team on human-machine interaction within the Institute of High Performance Computing on novel technologies for aerospace, manufacturing and retail.

EXPERTISE:
- Smart sensors
- Robotics
- AI

DR. NIKOLAI HLUBEK, BÜRKERT FLUID CONTROL SYSTEMS
EXPERT

Nikolai works as a Senior Data Scientist at Bürkert Fluid Control Systems and develops new data-driven products. He has a PhD in physics and has been using data science for more than 15 years to tackle various topics. As an example, before joining Bürkert Fluid Control Systems, he developed a real-time ionospheric monitoring service for the German Aerospace Center.

EXPERTISE:
- Data Science
- Product development
- Methodology

DANIEL BURKHARDT, FERDINAND-STEINBEIS-INSTITUT
EXPERT

Daniel is a researcher at the Ferdinand Steinbeis Institute with focus on the design of data-driven solution. Discovering patterns in data by the means of new technologies will tell us more about society, culture, and the environment. Mined properly, we will receive valuable insights that support us to create better, more precise and trustful solutions. While following this vision, Daniel aims at creating a tool that supports companies to design such innovative solutions. In this context, he leads with great motivation the AIoT Lab in Heilbronn, Germany and the IIC German Regional Team with the goal to transfer research findings into practical applications. He draws his inspiration from experiences during his Master study in Information Systems, Trainee program at Bosch GmbH and numerous digitalisation projects at the Ferdinand Steinbeis Institute.

EXPERTISE:
- Information Systems
- Digital transformation

HARINDERPAL HANSPAL, MOMENTA
EXPERT

Harinderpal (Hans) Hanspal is a Venture Partner at Momenta, a venture firm investing in industrial startups at the intersection of industry and digital. He has 25+ years of experience as a technology startup founder, corporate executive, entrepreneur, and, more recently, a venture capital investor driving business and product transformations in the technology, industrial, and telecom industries. He has held sales, product, and corporate strategy leadership roles at GE, Pivotal Software, VMware, and EMC (now DellEMC). As co-founder and COO of Nurego, an Industrial IoT Monetization startup, he helped grow from conception to a successful exit to GE. In his spare time, he leads Seattle's 4200+ member IoT Hub Meetup group.

EXPERTISE:
- Corporate Innovation
- Industrial Startups

Expert Network

JUAN GARCIA, MM1
EXPERT

Juan Sebastian Trujillo Garcia is a senior consultant at mm1. During his studies in economics he collected experience in business development, IT, and consulting trough different internships for enterprises like Daimler AG and Deutsche Telekom. For three years he has been working as a consultant for mm1.

EXPERTISE:
- Data Analysis
- Data Science

ALEXANDER WILLINEK, EVACO GMBH
EXPERT

More than 18 years of passion in BI projects – these are years in retrospect for Alexander Willinek, where he has gained a wealth of experience with a lot of passion and curiosity. As a programmer, he loves the situations where he develops solutions for customers in the first row, even if these are becoming increasingly rare in his current job as managing director and founder of EVACO GmbH.

EXPERTISE:
- Business Development
- Business Intelligence

ARNE FLICK, REPEATMOBILE
EXPERT

As CEO of repeatmobile, Arne has a strong focus on modernizing the employee learning experience and fostering transformational trainings. Together with his team, Arne develops digital solutions to significantly increase the learning transfer of trainings & seminars. Arne has over 20 years of experience in digital learning projects with national and international stock market listed companies. He is active in the Industrial IoT Ecosystem as forum speaker for Industry 4.0 of SIBB and DLT speaker of IoT+[Network] and formed educational joint-ventures with the IOTA Foundation, IoT ONE and DroneMasters under the name "IoT ONE Academy".

EXPERTISE:
- Digital Learning
- DLT
- IIoT

BENJAMIN LINNIK, OPITZ CONSULTING
EXPERT

Benjamin Linnik is a senior data engineer at Opitz Consulting. During his academic career, he has acquired profound knowledge in mathematics, physics and computer science. He uses his analytical skills to precisely grasp complex business and technical issues and to design customer-oriented IT solutions. During his Ph.D. Benjamin contributed to the development of a particle detector for a heavy-ion experiment at the international particle accelerator facility FAIR. For this purpose, he evaluated data statistically in order to ensure the suitability of the used components. Benjamin also used to automate complex (laboratory) workflows.

EXPERTISE:
- Machine Learning
- Computer Science
- Programming

CHRISTIAN WEISS, HOLISTICON AG
EXPERT

As a consultant, coach and trainer, Christian deals with the topics of business process management and agile project management. In particular, it is important to him to support large companies in the introduction of nimble, automated business processes and agile practices. Social concerns often play a major role in the implementation of ideas, for which he has developed a sensitive sense and sensitivity over the years.

EXPERTISE:
- UML
- BPMN

CHRISTOPH VOIGT, NEXOCRAFT GmbH
EXPERT

Christoph Voigt is the IoT & AI Solution Manager at nexocraft GmbH, where he offers customised solutions for those who want to delve into the world of AI. He is a graduate engineer in electrical engineering, who, having discovered AI as a new basis for modern automation technology in recent years, works closely with customers to implement the new AI Controller and offers client-specific solutions.

EXPERTISE:
- Machine Learning
- Automation technology

DANIEL WERTH, FERDINAND-STEINBEIS-INSTITUT
EXPERT

Daniel Werth is Senior Researcher and Director Multilateral Ecosystems at the Ferdinand Steinbeis Institute. In transfer-oriented projects in digital-based ecosystems, he strives to generate new added value for companies and added value to society. Daniel has over 15 years of experience in medium-sized companies, with a focus on wholesale and new services/business transformation. He received his doctorate from the LMU Munich in the field of business psychology.

EXPERTISE:
- Business Transformation
- Multilateral Ecosystems

PROF. DR. DIETLAND ZÜHLKE, TH KÖLN
EXPERT

Prof. Dr. Dietlind Zühlke is a professor for applied mathematics at the TH Köln since 2019. Before Dietlind worked at the TH Köln, she was a data science manager at Horn & Company Data Analytics. She was primarily responsible for projects for the conception and implementation of data science and machine learning applications, for the development of machine learning competences as well as programs for the transformation towards data-driven companies. Dietlind studied Computer

Science with a focus on Artificial Neural Networks at the Universities of Leipzig and Bonn. She earned a PhD in Computational Intelligence from the University of Groningen, Netherlands.

EXPERTISE:
- Data mining
- Machine Learning
- Algorithms

DIRK CASPER, OPITZ CONSULTING
EXPERT

Dirks motivation is to develop and implement viable and sustainable solutions together with customers and partners. From the brainstorming, development, project management and sales of IT solutions as well as change facilitation, he has worked through all roles. Together with connectivity partners and sensor manufacturers, he designed solutions for customers, identified suitable IOT platforms and helped developing complete ecosystems. At Opitz Consulting he is part of the AI core team and responsible for NLP, NLU, chat and voice bots. This also in combination with technologies such as augmented reality.

EXPERTISE:
- AIoT
- NLP
- NLU

ERIK WALENZA-SLABE, IOT ONE
EXPERT

As CEO of IoT ONE, Erik researches the impact of industrial digitalization on his client's businesses and supports them to define and implement their digitalization strategies. He has worked in China for eleven years where he is an active member of the Chinese innovation ecosystem. Aside from his role at IoT ONE, Erik is chair of the Technology and Innovation Committee at the American Chamber of Commerce, and Shanghai Director of Startup Grind, the world's largest community of entrepreneurs.

EXPERTISE:
- Industrial digitalization

FERMIN FERNANDEZ, ROBERT BOSCH TOOL CORPORATION
EXPERT

Fermin Fernandez is Director of Innovation at Bosch U.S. and Managing Director of the Chicago Connectory, an innovation center with a vibrant community of entrepreneurs and experts around IoT technologies and business ideas. Fermin has been with Bosch for 20+ years and has led technology projects in many locations around the world and diverse business units. He is an assessor for internal project manager certifications and holds an MBA from Wayne State University and an Industrial Engineering Bachelor degree.

EXPERTISE:
- IoT
- Industrial Engineering

GENE WANG, BOSCH.IO CHINA
EXPERT

Strategic Consultant of IoT & Digital Transformation. Gene Wang, the General Manager of Bosch.IO China, has been active in IoT market for more than 10 years, and provides consultant services and helps customers on their digital transformation strategy and new IoT business model with innovative technologies and solutions. Besides, Gene has more than 25 years' experience in Energy Area, and he held various positions at GE Energy.

EXPERTISE:
- IoT
- Smart Energy
- Industry 4.0

HAJO NORMANN, ACCENTURE
EXPERT

Hajo's interest in business-focused, enterprise-wide, cross siloed bundles of functionality arose in 2001 while he was responsible for a shared service platform at a large German bank as an architect and technical team lead. For many years he is helping to motivate, designing and implementing solutions successfully at various customers and work on choosing the right mix of tools, on setting up a successful modernization, sourcing & vendor strategy, and sharing Integration Architecture principles, design guidelines, and best practices. Haj has worked as a consultant, sales consultant, management consultant, technical team lead, and architect in large project teams in banking, retail, government, and telcos.

EXPERTISE:
- BPM
- Methodology
- SOA Design Pattern

HEINER DUFFING, ROBERT BOSCH GMBH
EXPERT

Heiner has more than 25 years' experience in purchasing and partially business development in various business areas (Steel, Automotive, Consumer, Renewables) and countries. Strong focus has been to find market innovations and develop start-up suppliers/products to reliable serial partners, including the negotiation of fitting contracts. Currently he leads the Purchasing of Software and Engineering Services for Bosch products. He holds a degree as Diplom-Wirtschaftsingenieur from TU Darmstadt.

EXPERTISE:
- Business development
- Purchase
- Logistics

DR. HOLGER KENN, MICROSOFT
EXPERT

As Director of Business Strategy in Microsoft's Business Development, Strategy and Ventures organization, Holger is responsible for defining and implementing strategy and investments in artificial intelligence, mixed reality and silicon ranging from tiny edge devices to global datacenter networks. Holger also represents Microsoft in industry bodies such as the OPC Foundation, the Industrial Internet Consortium, and the Digital Twins Consortium. Before joining Microsoft, Holger held several academic positions in wearable computing, robotics and AI research. He holds a Ph.D. in computer science and has more than 20 years of experience in artificial intelligence and the software industry.

EXPERTISE:
- Strategy and Ventures
- Business Development

JIM MORRISH, FOUNDING MEMBER, TRANSFORMA INSIGHTS
EXPERT

Jim is a respected Digital Transformation and Internet of Things industry expert, with over 20 years' experience of strategy consulting, operations management and telecoms research. Previously he was a Founder and the Chief Research Officer of Machina Research, the world's leading IoT analyst firm, which was acquired by Gartner in 2016. He is a co-author of the Ignite IoT framework on which the Ignite AIoT framework is based.

EXPERTISE:
- AIoT Business Models

JÖRN EDLICH, MM1
EXPERT

Jörn Edlich is a senior manager at mm1. He is responsible on digital, IoT-based services in the mobility and energy industry and focuses on the IT-security aspects of IoT-architectures. He graduated with a diploma degree in Electrical Engineering and Communication Technology at RWTH Aachen University in 2006.

EXPERTISE:
- Telecommunications
- Business Development

KAI HACKBARTH, BOSCH.IO
EXPERT

Kai Hackbarth is Business Owner Industrial at Bosch.IO, and also co-chair of the OTA SIG at the Industrial Internet Consortium. Kai has many years' experience with IoT applications and architectures. He also has served as a Director of the OSGi Alliance for many years.

EXPERTISE:
- OTA
- Security

DR. KATHARINA MATTES, MM1
EXPERT

Dr. Katharina Mattes is a manager at mm1. After her diploma in business economics and her Ph.D. in innovation management she worked for VDMA, as head of coordination office. VDMA is a mechanical engineering industry association. Her work was about Industry 4.0 Baden-Wuerttemberg and gave her over 3 years of experience in Industry 4.0. In 2020 she also did her professional scrum master and started working for mm1.

EXPERTISE:
- Industry 4.0
- Business Engineering

KIM KORDEL, BOSCH.IO
EXPERT

Kim Kordel is a senior business development manager for new IoT business at Bosch.IO. In her former position as an IoT business consultant and trainer for IoT business models at Bosch.IO she developed and taught methodology for building IoT business models. With this methodology she developed new digital business for internal and external customers. Kim also co-initiated and set-up the Bosch Startup Harbour, the incubation program for external startups for Bosch. Now Kim is responsible to establish new IoT business for the energy domain.

EXPERTISE:
- IoT Business
- Ecosystem Business Models

LAURENZ KIRCHNER, MM1
EXPERT

Laurenz Kirchner is a partner at mm1 with a consulting focus on driving digital growth, helping his clients to become truly connected businesses. As a trained architect, MBA and experienced management consultant (including eight years McKinsey), Laurenz is a hybrid thinker with a strong ability to combine creative problem solving with solid business judgement. His consulting projects center on innovation, the development of digital products and ecosystems as well as data strategies for the Internet of Things.

EXPERTISE:
- Driving Digital Growth
- Architecture

PROF. LIRONG ZHENG, FUDAN UNIVERSITY
EXPERT

Lirong Zheng received his Ph.D. degree from the Royal Institute of Technology (KTH), Stockholm, Sweden in 2001. Afterwards he worked at KTH as a research fellow, associate professor and full professor, expert of Ericsson etc. He is a distinguished professor since 2010 at Fudan University, Shanghai China. Currently, he holds the directorship of Shanghai Institute of Intelligent Electronics and Systems, Fudan University. His research experience and interest includes ambient intelligence and internet-of-things, and applications in industry and Fintech etc. He has authored more than 400 publications and servers as steering board member of International Conference on Internet-of-Things.

EXPERTISE:
- AIoT
- IoT and Edge Computing
- BlockChain

MARCUS SCHUSTER, ROBERT BOSCH GMBH
EXPERT

As a Project Director for embedded AI Marcus Schuster runs a cross-cutting project within the Robert Bosch Group, which creates prototypic realizations of products with embedded AI based core functions. He leads a diverse team of AI experts, embedded and backend SW engineers, architects, hardware developers and project managers. Marcus did his PhD on superconducting electronics, and joined Bosch in 2005. Since then he held various management positions in quality, business development and both HW and SW engineering.

EXPERTISE:
- Embedded AI

MARC HABERLAND, CLARIBA
EXPERT

Marc has more than 18 years' experience in Business Intelligence (BI), analytics, strategy management and Enterprise Performance Management (EPM) across telecommunication, education, healthcare, manufacturing, banking and public sectors. Marc leads a team of 70+ BI and analytics experts who deliver innovative, reliable and high-quality data driven digital transformation solutions, providing customers with clarity and actionable insight to improve business performance and develop new business models for long-term, sustainable competitive advantage.

EXPERTISE:
- BI
- EPM

MICHAEL HOHMANN, BSH HAUSGERÄTE GMBH
EXPERT

Dr. Michael Hohmann works as a Systems Engineer in the field of autonomous cleaning robots at BSH. After studies in Mechanical Engineering and Measurement Science at TU Ilmenau he joined the Bosch Roxxter development team at BSH robotics department.

EXPERTISE:
- Robotics and Architecture

DR. MICHAEL WENIGER, DEUTSCHEPOST DHL GROUP
EXPERT

Dr. Michael Weniger is a Senior Data Scientist at Deutsche Post DHL Group. During his time at Deutsche Post he collected over 4 years of experience in Artificial Intelligence and Deep Learning. Michael graduated with a Diploma in Mathematics and holds a PhD in probability-theoretical weather and climate models.

EXPERTISE:
- Machine Learning
- AI
- Deep Learning

PABLO ENDRES, SEVENSHIFT
EXPERT

Pablo Endres, Founder of SevenShift GmbH. Experienced security consultant and Professional Hacker. Pablo's career has taken place mostly doing security in a variety of industries, like Cloud Service providers, Banks, Telecommunications, contact centers, and universities. He holds a degree in computer engineering, as well as a handful security certifications. Pablo has founded multiple companies in different continents and enjoys hacking, IoT, teaching, working with new technologies, start-ups, collaborating with Open Source projects and being challenged.

EXPERTISE:
- IoT Security
- Management of IT Projects

PETER KLEMENT, AVANADE DEUTSCHLAND GMBH
EXPERT

As Digital Enterprise Advisory Lead, Peter is responsible for helping clients to leverage digital innovations to create business value following the Triple Bottomline concept. Before joining Avanade, Peter built the IoT Practice for DXC.technology in Australia and New Zealand and was active for many years in the Industrial

Internet Consortium. He is also Vice President of the MIT Club of Germany, where he is responsible for partnerships and programs. Peter das a MS in Computer Science for the University of Applied Science in Munich and an MBA degree from the MIT Sloan School of Management.

EXPERTISE:
- IoT

PETER LINDLAU, TOMORROW LABS GMBH
EXPERT

Peter Lindlau is working for over 4 years at Tomorrow Labs as an operative partner and has many years of experience with the implementation of IT projects. His current focus (for about six years) is the implementation of Industry 4.0 projects. Peter is the project manager of the BMBF funding project "eApps4Production".

EXPERTISE:
- Business Development
- Business Strategy

DR. ROBERT XIE, BOSCH COOPERATE RESEARCH
EXPERT

Research Expert of IoT Applications Head of IoT@Life program and group leader for IoT & I4.0 at the Bosch Research and Technology Center in China. Before joining Bosch, Robert was an Assistant Professor at Shanghai Jiao Tong University focusing on medical robotics and sensor systems. As an alumni of the Bosch Accelerator Program, he not only collaborates closely with Business Units for joint development, but also explore commercialization of IoT products and solutions with scalable, repeatable and profitable business models.

EXPERTISE:
- IoT
- Sensor Systems
- Computer Visions

PROF. DR. THOMAS BARTZ-BEIELSTEIN, TH KÖLN
EXPERT

Director of the Institute for "Data Science, Engineering and Analytics" and Professor for Applied Mathematics at TH Köln. More than 20 years' experience in simulation, optimization, machine learning and AI.

EXPERTISE:
- Machine Learning
- Modelling and Simulation

SANGAMITHRA PANNEER SELVAM, FERDINAND-STEINBEIS-INSTITUT
EXPERT

Sangamithra is a working student at Ferdinand Steinbeis Institut. She has one year of experience working on AIoT framework and AIoT laboratory: Pneumatic systems use case. After her bachelor's degree, she moved to Germany to pursue her masters in Electrical Engineering, majoring in Smart Information Processing at the University of Stuttgart.

EXPERTISE:
- AIoT
- Data Science

STEPHAN WOHLFAHRT, ROBERT BOSCH GMBH
EXPERT

As Director and Corporate Process Owner Project Management at Bosch, Stephan is responsible for the general project management guidelines and company-wide PM qualification offerings of the Bosch Group, including predictive/plan driven, agile, and hybrid approaches. He has more than 20 years of experience in project management and organizational development and is a representative in PMI's Global Executive Council.

EXPERTISE:
- Project management
- Agile planning

THOMAS JAKOB, DRAEGER
EXPERT

As Chief Operating Officer for the Africa, Asia, and Australia region, Thomas supports its customers and the organization to scale efficiently, leveraging leading-edge technologies and digitalization also to drive transition to the new economy. For close to nine years, he was before spearheading the development of AIoT solutions and business models as Regional President of Bosch.IO in Asia Pacific. With master degrees in both electrical engineering and business administration, Thomas has more than 20 years of experience in senior management positions in industries such as IT and telecommunications, manufacturing and management consulting.

EXPERTISE:
- IT
- Telecommunications
- Digitalization

TORSTEN WINTERBERG, INNOVATION HUB BERGISCHES RHEINLAND e.V./OPITZ CONSULTING Deutschland GMBH
EXPERT

Torsten Winterberg holds degrees in electrical engineering, computer science and business administration. As Director of Business- and IT-Innovation he is responsible for future topics at OPITZ CONSULTING. Torsten is a consultant and coach for "everything new", his focus is on software-driven solutions, AI, IoT, innovative capabilities, digital business models, solution concepts and architectures. In addition, Torsten is actively developing the Innovation Hub Bergisches RheinLand e.V. as managing director.

EXPERTISE:
- AIoT
- Architecture
- Digital Business Models

DR. ACHIM NONNENMACHER, ROBERT BOSCH GMBH
EXPERT

Achim's goal is to improve the experience of vehicle users, OEMs, and software developers by accelerating the innovation cycles in the automotive industry. At Bosch Connected Mobility Solutions he is responsible for product and portfolio of the Software-defined Vehicle. Before this, he drove product innovations in the mobility sector by validating business, technology hypotheses and user needs at scale. Achim holds a PhD in computational mathematics from Swiss Institute of Technology (EPFL) and an executive education on Innovation Acceleration from UC Berkeley.

EXPERTISE:
- Automotive
- Product development
- Business Strategy

HANNAH ABELEIN, MM1
EXPERT

Dr. Hannah Abelein is a Manager for Research and Knowledge Management at mm1. During her PhD thesis in software engineering, she gained experience in the fields of managing large-scale IT Projects in various industries and studied how the communication between the various roles can improve project success. As a consulted she work for various clients mainly in the digitalization area, such as DHL, Deutsche Bank, Adidas and Thomson Reuters. In addition, she founded and is the content lead for the AI Circle – a network for practical exchange on AI Projects

EXPERTISE:
- Consultant

References

1. *Start with Why: How great leaders inspire everyone to take action*, Simon Sinek, 2010
2. *The Business Model Navigator: 55 Models That Will Revolutionise Your Business*, Oliver Gassmann, Karolin Frankenberger, Michaela Csik, 2014
3. Russell, Stuart J.; Norvig, Peter (2003), Artificial Intelligence: A Modern Approach (2nd ed.), Upper Saddle River, New Jersey: Prentice Hall, ISBN 0-13-790395-2
4. Mitchell, Tom (1997). Machine Learning. New York: McGraw Hill. ISBN 0-07-042807-7. OCLC 36417892.
5. *Conceptual Approaches for Defining Data, Information, and Knowledge*, Zins, Chaim, 2007, Journal of the American Society for Information Science and Technology.
6. *From model-centric to data-centric*, Fabiana Clemente, 2021, https://towardsdatascience.com/from-model-centric-to-data-centric-4beb8ef50475
7. *Platform Industrie 4.0: Asset Administration Shell*, https://www.plattform-i40.de/PI40/Redaktion/EN/Downloads/Publikation/vws-in-detail-presentation.pdf
8. *Digital Twin Consortium: The Definition of a Digital Twin*, https://www.digitaltwinconsortium.org/initiatives/the-definition-of-a-digital-twin.htm
9. *Digital Twins Definition Language (DTDL)*, https://docs.microsoft.com/en-us/azure/digital-twins/how-to-manage-model
10. *Platform Revolution: How Networked Markets Are Transforming the Economy - And How to Make Them Work for You*, Geoffrey G. Parker, Marshall Van Alstyne, and Sangeet Paul Choudary, 2016, Norton & Company
11. *How Smart, Connected Products Are Transforming Competition*, Michael Porter and Jim Heppelmann, 2014, Harvard Business Review
12. *Crossing the Chasm*, Geoffrey Moore, 1991, HarperBusiness
13. *Business Model Generation*, Alexander Osterwalder, Yves Pigneur, Alan Smith, and 470 practitioners from 45 countries, self-published, 2010
14. *The Business Model Navigator: 55 Models That Will Revolutionise Your Business*, Oliver Gassmann, Karolin Frankenberger, Michaela Csik, 2014
15. *Distribution of Cost over the Application Lifecycle - a Multi-case Study*, Ruediger Zarnekow, Walter Brenner, 2005
16. *FSM Definition*, Gartner Group, 2019
17. *ITIL Service Operation Processes Explained*, A. Brahmachary, 2018
18. *Definition: CMDB - Configuration Management Database*, M. Rouse, 2017
19. *The Essential Guide to Creating an AI Product in 2020*, Rahul Parundekar, 2020
20. *Machine learning for streaming data: state of the art, challenges, and opportunities*, H. Gomes et. al., 2020

© AIoT User Group 2023
D. Slama et al. (eds.), *The Digital Playbook*,
https://doi.org/10.1007/978-3-030-88221-1

21. *Strategic management and organisational dynamics: the challenge of complexity. 3rd ed.*, Stacey RD., Prentice Hall, 2002
22. *Disciplined Agile*, https://pmi.org/disciplined-agile
23. *The Connected Company*, Dave Gray, Thomas Vander Wal, O'Reilly Media, Inc., 2012
24. *Site Reliability Engineering: How Google Runs Production Systems*, N. Murphy et al., 2016
25. Chaos Engineering, Wikipedia
26. R3E – An Approach to Robustness, Reliability, Resilience and Elasticity Engineering for End-to-End Machine Learning, Hong-Linh Truong, 2020
27. Hidden Technical Debt in Machine Learning Systems, D. Sculley et al., 2015
28. *Data Readiness: Using the "Right" Data*, Alex Castrounis, 2010
29. *FAIR Guiding Principles for scientific data management and stewardship*, GO FAIR, Scientific Data, 2016, https://www.go-fair.org/fair-principle
30. *Why Jupyter is data scientists' computational notebook of choice*, Nature 563, 145-146 (2018), https://doi.org/10.1038/d41586-018-07196-1

Printed in the United States
by Baker & Taylor Publisher Services